TEMPERATURES
VERY LOW
and
VERY HIGH

TEMPERATURES
VERY LOW
and
VERY HIGH

MARK W. ZEMANSKY

Professor Emeritus of Physics
The City College of New York

Dover Publications, Inc.
New York

TO MY SONS,
PHILIP AND HERBERT

Published in Canada by General Publishing Company,
Ltd., 30 Lesmill Road, Don Mills, Toronto, Ontario.
Published in the United Kingdom by Constable and
Company, Ltd., 10 Orange Street, London WC2H 7EG.

This Dover edition, first published in 1981, is an
unabridged and corrected republication of the work
originally published for The Commission on College
Physics, under the general editorship of Edward U. Condon,
by D. Van Nostrand Company, Inc., in 1964.

International Standard Book Number: 0-486-24072-X
Library of Congress Catalog Card Number: 80-69673

Manufactured in the United States of America
Dover Publications, Inc.
180 Varick Street
New York, N.Y. 10014

Preface

The concept of temperature, like the concept of force, makes its entrance into physics through the door of bodily sensation, or sense perception. Just as we think of force in terms of muscular exertion, we think of temperature in terms of hotness or coldness. The concept of force, however, must be freed from the human body if it is to be applied to the attraction of the earth to the sun, or the attraction of an electron to a proton. By the same token, the concept of temperature must be divorced from bodily sensation to be capable of describing paramagnetic salts that have been demagnetized, or gases supporting a shock wave. How this is done, how temperature is measured, and how extremes of temperature are produced and utilized are described in this book.

It has been written for anyone who is beginning the serious study of science—such as a high school or college student who is taking a first-year physics course and who plans to major in one of the sciences. It is intended to give him more insight into the meaning and importance of temperature than is provided by the usual textbook of college physics. The first two chapters are devoted to the concept of temperature in macroscopic physics and in statistical or microscopic physics. The production and measurement of temperatures near absolute zero are then described in considerable detail. This is followed by a chapter on the production and measurement of high temperatures up to the fifty-million-degree range. The last chapter goes beyond infinity into the realm of negative temperatures.

The author would like to express his thanks to Louis A. Turner and to Edward U. Condon for their patience and kindness in suggesting valuable alterations and corrections. Almost all of these suggestions were followed.

M. W. Z.

Teaneck, N.J.

Table of Contents

Preface   v

1 Temperature Is a Property of Matter   1
   *The Concept of Temperature, 1; Thermometers, 5; The Es-
   tablishment of a Temperature Scale, 9; The International
   Temperature Scale, 15; Heat, a Form of Energy, 16; Isothermal and Adiabatic Processes, 18; The Kelvin Temperature Scale, 21*

2 Temperature, Entropy, and Disorder   25
   Energy and Entropy, 25; The Molecular Point of View, 29

3 The Approach to Absolute Zero   37
   *The Joule-Kelvin Effect, 37; In the Liquid Helium Region,
   41; Paramagnetic Salts, 46; Adiabatic Demagnetization, 50;
   Conversion of Magnetic Temperature to Kelvin Tempera-
   ture, 55; Superconductivity, 58; Magnetic Refrigerator, 64;
   The Polarization of Magnetic Nuclei, 64; Production of Still
   Lower Temperatures by Nuclear Demagnetization, 69; The
   Third Law of Thermodynamics, 73*

4 The Approach to Infinite Temperature   77
   *Filaments, Furnaces, and Flames, 77; Planck's Radiation Law
   and the Optical Pyrometer, 79; Arcs and Plasmas, 86; Spectral
   Lines, 90; Shock Waves, 100; Fission of Uranium, 105; Fusion
   Reactions, 107*

5 Beyond Infinity to Negative Temperatures   112
   *Subsystems, 112; Negative Values of the Kelvin Tempera-
   ture, 114; Achievement of Negative Temperatures, 118;
   Thermodynamics at Negative Temperatures, 120*

   Bibliography   123

   Index   125

Plates (following p. 88)

I   The famous electromagnet of the Laboratoire Aimé Cotton at Bellevue, near Paris

II   Apparatus used at Oxford for the production of temperatures in the microdegree range

III   Plasma jet in nitrogen

IV   Mach-Zehnder fringes, showing three stages in the formation of a shock wave

TEMPERATURES
VERY LOW
and
VERY HIGH

1. Temperature Is a Property of Matter

§ 1-1 **The concept of temperature.** Ever since early childhood we have experienced the sensations of hotness and coldness and have described these sensations with the aid of adjectives such as cold, cool, tepid, warm, hot, etc. When we touch an object, we use our *temperature sense* to ascribe *to the object* a property called *temperature* which determines whether it feels hot or cold to the touch. The hotter it feels, the higher the temperature. This procedure plays the same role in "qualitative science" that hefting a body does in determining its weight or that kicking an object does in estimating its mass. To determine the mass of an object, we must first arrive at the concept of mass by means of *quantitative* operations with the aid of an instrument like an inertia balance or an equal-arm balance. This set of operations is performed without appeal to the sense perceptions associated with flexed muscles or to the discomfort connected with kicking. Similarly, the quantitative determination of temperature requires a set of operations that are independent of sense perceptions of hotness or coldness and that involve measurable quantities. How this is done is explained in the following pages.

The state of certain simple systems may be specified by measuring the value of one physical quantity. Consider, for example, a liquid such as mercury or alcohol contained in a very thin-walled bulb which communicates with a very narrow tube or capillary, as shown in Fig. 1-1(a). The state of this system is specified by noting the length of the liquid column, starting at an arbitrarily chosen point. The length L is called a *state coordinate*. Another simple system is shown in Fig. 1-1(b), which depicts a thin-walled vessel containing a gas whose volume remains con-

1

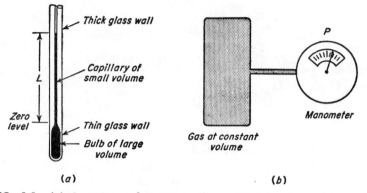

**FIG. 1-1   (a) A system whose state is specified by the value of L.
(b) A system whose state is given by the value of P.**

stant and whose state coordinate is the pressure P, as read on any convenient pressure gauge. In the next section use is made of a coil of fine wire (at constant tension) whose state coordinate is the value of its electrical resistance, and also a junction of two dissimilar metals whose state coordinate is the value of the electromotive force between its ends.

Let A stand for the liquid-in-capillary system with state coordinate L and let B stand for the gas at constant volume with state coordinate P. If A and B are brought into contact, their state coordinates, in general, are found to change. When A and B are separated, however, the change is slower, and when thick walls of various materials such as wood, plaster, felt, or asbestos are used to separate A and B, the values of the respective state coordinates L and P are almost independent of each other. Generalizing from these observations, we postulate the existence of an ideal partition, called an **adiabatic wall,** *which, when used to separate two systems, allows their state coordinates to vary over a large range of values* **independently.** An adiabatic wall is an idealization that cannot be realized perfectly, but may be approximated closely. It is represented as a thick cross-shaded region, as in Fig. 1-2(a).

When A and B are put into actual contact or are separated by a thin metallic partition, their state coordinates change. *A wall that enables one system to influence another (as shown by changes in the state coordinates of both) is called a* **diathermic wall.** A thin sheet of copper is the most practical diathermic wall. In Fig. 1-2(b), a diathermic wall is depicted as a thin, darkly

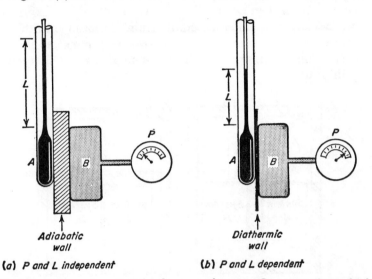

Adiabatic wall

Diathermic wall

(a) *P and L independent*

(b) *P and L dependent*

FIG. 1-2 System A, a liquid column, and system B, a gas at constant volume, separated by (a) an adiabatic wall and (b) a diathermic wall.

shaded region. When systems *A* and *B* are in contact through a diathermic wall, their coordinates *L* and *P* change until after a time no further change takes place. *When two systems are separated by a diathermic wall, the joint state of both systems that exists when all changes in the coordinates have ceased is called* **thermal equilibrium.**

Imagine two systems *A* and *B* separated from each other by an adiabatic wall, but each in contact with a third system *C* through diathermic walls, the whole assembly being surrounded by an adiabatic wall as shown in Fig. 1-3(a). Experiment shows that the two systems come to thermal equilibrium with the third and that no further change occurs if the adiabatic wall separating *A* and *B* is then replaced by a diathermic wall (Fig. 1-3(b)). If, instead of allowing both systems *A* and *B* to come to equilibrium with C at the same time, we first have equilibrium between A and C and then equilibrium between B and C (the state of system C being the same in both cases), then, when A and B are brought into communication through a diathermic wall, they are found to be in thermal equilibrium. We shall use the expression "two systems are in thermal equilibrium" to mean that the two systems are in states such that, if the two *were* connected through a diathermic wall, the combined system *would be* in thermal equilibrium.

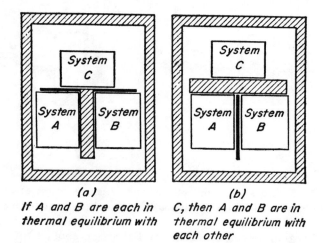

(a)   (b)

If A and B are each in   C, then A and B are in
thermal equilibrium with   thermal equilibrium with
                                            each other

FIG. 1-3   The zeroth law of thermodynamics.

These experimental facts may then be stated concisely in the following form: *Two systems in thermal equilibrium with a third are in thermal equilibrium with each other.* Following R. H. Fowler, we shall call this postulate the *zeroth law of thermodynamics.* At first it might seem that the zeroth law is obvious, but this is not so. An amber rod A that has been rubbed with fur will attract a neutral pith ball C. So will another similarly rubbed amber rod B, but the two amber rods will not attract each other. (Woman A loves man C. Woman B loves man C. But does woman A love woman B?)

When two systems A and B are first put in contact through a diathermic wall, they may or may not be in thermal equilibrium. One is entitled to ask, "What is there about A and B that determines whether or not they are in thermal equilibrium?" Experiment shows that neither the mass, the density, the elastic modulus, the electric charge, nor the magnetic state—in fact, none of the quantities of importance in mechanics, electricity, or magnetism—is the determining factor. *We therefore* **infer** *the existence of a new property called the* **temperature.** *The temperature of a system is that property which determines whether or not it will be in thermal equilibrium with other systems. When two or more systems are in thermal equilibrium, they are said to have the same temperature.*

The temperature of all systems in thermal equilibrium may be represented by a number. The establishment of a temperature scale is merely the adoption of a set of rules for assigning numbers to temperatures. Once this is done, the condition for thermal equilibrium between two systems is that they have the same temperature. Also, when the temperatures are different, we may be sure that the systems are not in thermal equilibrium.

The preceding operational treatment of the concept of temperature emphasizes the fundamental idea that the temperature of a system is a property which eventually attains the same value as that of other systems when all these systems are put in contact or separated by thin metallic walls within an enclosure of thick asbestos walls. This concept is identical with the everyday idea of temperature as a measure of the hotness or coldness of a system, since, so far as our senses may be relied upon, the hotness of all objects becomes the same after they have been together long enough. However, it is necessary to express this simple idea in technical language in order to be able to establish a rational set of rules for measuring temperature and also to provide a solid foundation for the study of thermodynamics and statistical mechanics.

§1-2 **Thermometers.** To determine the temperatures of a number of systems, the simplest procedure is to choose one of the systems arbitrarily as an indicator of thermal equilibrium between it and the other systems. The system so chosen is called a *thermometer,* and its state coordinate is called its *thermometric* *property*. The zeroth law provides that the reading of the thermometer is the temperature of *all* systems in thermal equilibrium with it. The important characteristics of a thermometer are *sensitivity* (an appreciable change in the thermometric property produced by a small change in temperature), *accuracy* in the measurement of the thermometric property, and *reproducibility*. Another often desirable property is *speed* in coming to thermal equilibrium with other systems. The thermometers which satisfy these requirements best are described in the following paragraphs.

The most useful thermometer in most research and engineering laboratories is the *thermocouple*, which consists of a junction of two different metals or alloys, and which is labeled "test junction" in Fig. 1-4. The test junction is usually imbedded in the

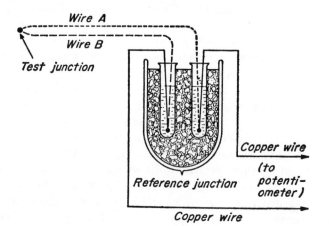

FIG. 1-4   A thermocouple, showing the test junction and the reference junction.

material whose temperature is to be measured. Since the test junction is small and has a small mass, it can follow temperature changes rapidly and come to equilibrium quickly. The reference junction consists of two junctions: one of A and copper and the other of B and copper. These two junctions are maintained at any desired constant temperature, called the reference temperature. The thermometric property of this thermometer is an electrical quantity called the emf ("ee-em-eff," electromotive force) which is measured with an instrument known as a potentiometer. A thermocouple with one junction of pure platinum and the other of 90 percent platinum and 10 percent rhodium is often used. Copper and an alloy called "constantan" are also frequently used.

A *resistance thermometer* consists of a fine wire, often of platinum, wound on a mica frame and enclosed in a thin-walled silver tube for protection. Copper wires lead from the thermometer unit to a resistance-measuring circuit such as a Wheatstone bridge. Since resistance may be measured with great precision, the resistance thermometer is one of the most precise instruments for the measurement of temperature. In the region of extremely low temperatures, a small carbon cylinder or a small piece of germanium crystal is often used instead of a coil of platinum wire.

To measure temperatures above the range of thermocouples and resistance thermometers an *optical pyrometer* is used. As shown in Fig. 1-5, it consists essentially of a telescope *T,* in the tube of which is mounted a filter *F* of red glass and a small elec-

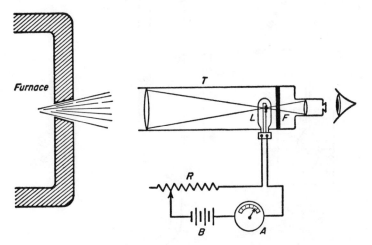

FIG. 1-5 Main features of an optical pyrometer.

tric lamp bulb *L*. When the pyrometer is directed toward a furnace, an observer looking through the telescope sees the dark lamp filament against the bright background of the furnace. The lamp filament is connected to a battery *B* and a rheostat *R*. By turning the rheostat knob the current in the filament, and hence its brightness, may be gradually increased until the brightness of the filament just matches the brightness of the background. From previous calibration of the instrument at known temperatures, the scale of the ammeter *A* in the circuit may be marked to read the unknown temperature directly. Since no part of the instrument need come into contact with the hot body, the optical pyrometer may be used at temperatures above the melting points of metals.

Of all the various *thermometric properties,* the pressure of a gas whose volume is kept constant stands out for its sensitivity, accuracy of measurement, and reproducibility. The constant-volume gas thermometer is illustrated schematically in Fig. 1-6. The materials, construction, and dimensions differ in various

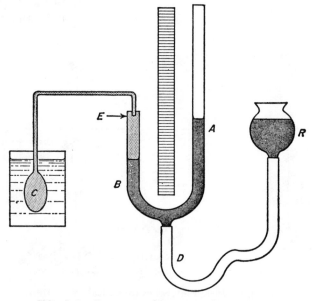

FIG. 1-6 Constant-volume gas thermometer.

laboratories throughout the world and depend on the nature of the gas and the temperature range to be covered.

The gas, usually helium, is contained in bulb *C*, and the pressure exerted by it is measured by the open-tube mercury manometer. As the temperature of the gas increases, the gas expands, forcing the mercury down in tube *B* and up in tube *A*. Tubes *A* and *B* communicate through a rubber tube *D* with a mercury reservoir *R*. By raising *R*, the mercury level in *B* may be brought back to a reference mark *E*. The gas is thus kept at constant volume.

Gas thermometers are used mainly in bureaus of standards and in some university research laboratories. They are usually large, bulky, and slow in coming to thermal equilibrium.

§1-3 **The establishment of a temperature scale.** Any one of the thermometers described in the preceding section may be used to indicate the constancy of a temperature if its thermometric property remains constant. By this means, it has been found that a system composed of a solid and a liquid of the same material maintained at constant pressure will remain in *phase equilibrium* (that is, the liquid and solid exist together without the liquid changing into solid or the solid changing into liquid) only at one constant temperature. Similarly, a liquid will remain in phase equilibrium with its vapor at only one definite temperature when the pressure is maintained constant.

The temperature at which a solid and liquid of the same material coexist in phase equilibrium *at atmospheric pressure* is called the *normal melting point,* abbreviated NMP.

The temperature at which a liquid and its vapor exist in phase equilibrium at atmospheric pressure is called the *normal boiling point,* abbreviated NBP.

Phase equilibrium between a solid and its vapor is sometimes possible at atmospheric pressure. The temperature at which this takes place is the *normal sublimation point,* NSP.

It is possible for all three phases—solid, liquid, and vapor—to coexist in equilibrium, but only at one definite pressure and temperature known as the *triple point,* abbreviated TP. The triple point pressure of water is 4.58 mm of mercury (Hg).

The NMP, NBP, NSP, or TP of any material can be chosen

as a standard for the purpose of setting up a temperature scale. Any temperature so chosen is called a *fixed point*. Before 1954 there were two standard fixed points, the NBP of water and the equilibrium temperature of pure ice and air-saturated water. Both of these have been abandoned. By international agreement, *there is now only one standard fixed point in modern thermometry, and that is the triple point of water,* to which is given the arbitrary number

$$273.16°K,$$

read "273.16 degrees Kelvin." The particular value is chosen so as to preserve the traditional difference of 100.00° between the NBP of water (373.15°) and the ice point (273.15°). Why Kelvin's

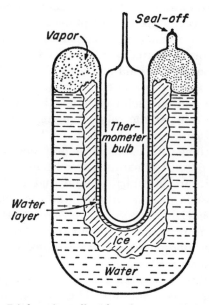

FIG. 1-7   Triple-point cell with a thermometer in the well.

name is used will be explained later. To achieve the triple point, water of the highest purity is distilled into a vessel like that shown schematically in Fig. 1-7. When all air has been removed, the vessel is sealed off. With the aid of a freezing mixture in the inner well, a layer of ice is formed around the well. When the freezing mixture is replaced by a thermometer bulb, a thin layer of ice is melted nearby. So long as the solid, liquid, and vapor phases coexist in equilibrium, the system is at the triple point.

We start our program of setting up a temperature scale by denoting with the letter X *any one* of the thermometric properties mentioned previously:

the emf of a thermocouple, $\mathcal{E}$

the resistance of a wire, R,

the pressure of a gas at constant volume, P, etc.

We define the ratio of two temperatures to be the same as the ratio of the two corresponding values of X. Thus, if a thermometer with thermometric property X is put in thermal equilibrium with a system and registers a value X, and it is then put in thermal equilibrium with an other system and registers a value X_3, the ratio of the temperatures of those two systems is given by

$$\frac{T(X)}{T(X_3)} = \frac{X}{X_3}. \tag{1-1}$$

If now, we let the subscript 3 stand for the standard fixed point, the triple point of water, then $T(X_3) = 273.16°\text{K}$. Hence,

$$T(X) = 273.16°\text{K}\,\frac{X}{X_3}. \tag{1-2}$$

It is important to understand that the relation represented by Eq. 1-1 is an *arbitrary choice*. The ratio of two temperatures could have been chosen to be the ratio of the squares of the X's, or the logarithm of the ratio of the X's, or the ratio of the negative reciprocals of the X's. Eq. 1-1 represents the *simplest* arbitrary choice.

The next step is to see what results are obtained when different thermometers are used to measure the same temperature, following rules laid down in Eq. 1-2. Table 1-1 lists the results of such a test in which four different thermometers—a copper-nickel thermocouple, a platinum resistance thermometer, and two hydrogen gas thermometers (one at high pressures and one at low pressures)—were used to measure the temperatures of six different fixed points. It is clear from Eq. 1-2 that the tempera-

TABLE 1-1 *Comparison of Thermometers*

Fixed point	(Cu-Ni) ε, mv	$T(\varepsilon)$	(Pt) R, ohms	$T(R)$	(H$_2$, const. V) P, atm	$T(P)$	(H$_2$, const. V) P, atm	$T(P)$
N$_2$ (NBP)	-0.10	-9.2	1.96	54.5	1.82	73	0.29	79
O$_2$ (NBP)	0	0	2.50	69.5	2.13	86	0.33	90
CO$_2$ (NSP)	$+1.52$	139	6.65	185	4.80	193	0.72	196
H$_2$O (TP)	$\varepsilon_3 = 2.98$	273	$R_3 = 9.83$	273	$P_3 = 6.80$	273	$P_3 = 1.00$	273
H$_2$O (NBP)	5.30	486	13.65	380	9.30	374	1.37	374
Sn (NMP)	9.02	826	18.56	516	12.70	510	1.85	505

ture of the triple point of water must come out to be 273.16°K. At any other temperature, however, Table 1-1 shows that the two gas thermometers agree quite well, but differ markedly from the readings of the other two thermometers. Further experiments show that the greatest agreement is found among gas thermometers and that, *regardless of the nature of the gas, all gas thermometers at the same temperature approach the same reading as the pressure of the gas approaches zero.* The results of an experiment of this kind are shown in Fig. 1-8, where four gas thermom-

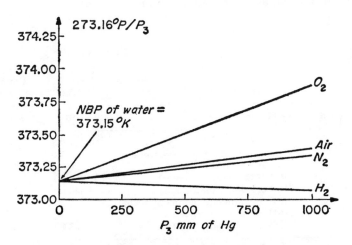

FIG. 1-8 Readings of four constant-volume gas thermometers at the normal boiling point of water, when different gases are used at various values of P_3.

eters give readings for the NBP of water which approach the same value, 373.15°K, as the pressure approaches zero.

The best temperature scale that we can devise at this time without further theoretical reasoning is the one that is least dependent on the particular properties of some single substance. It is the *ideal gas temperature T,* where

$$T = 273.16°\text{K} \lim_{P_3 \to 0} \frac{P}{P_3}, \tag{1-3}$$

and the symbolism on the right means the limiting value of the ratio P/P_3 as P_3 (and P along with it) approaches zero. The accurate measurement of an ideal gas temperature of some fixed point, such as the NBP of sulfur, requires months of painstaking laboratory work and mathematical computation and, when completed, is an international event. Such work is published in a physical journal and eventually listed in tables of physical constants. The temperatures of the normal boiling points (NBP) and normal melting points (NMP) of a number of materials have been measured, and the results are tabulated in Table 1-2. The fixed points designated in the table as basic are used to calibrate

TABLE 1-2   *Temperatures of Fixed Points*

	Fixed points	Temp., °C	Temp., °K
Standard	**Triple point of water**	**0.01**	**273.16**
Basic	NBP of oxygen (oxygen point)	−182.97	90.18
	Equilibrium of ice and air-saturated water (ice point)	0.00	273.15
	NBP of water (steam point)	100.00	373.15
	NBP of sulfur (sulfur point)	444.60	717.75
	NMP of antimony (antimony point)	630.50	903.65
	NMP of silver (silver point)	960.80	1233.95
	NMP of gold (gold point)	1063.00	1336.15
Secondary	NBP of helium	−268.93	4.22
	NBP of hydrogen	−252.78	20.37
	NBP of neon	−246.09	27.06
	NBP of nitrogen	−195.81	77.34
	NMP of mercury	−38.86	234.29
	Transition point of sodium sulfate	32.38	305.53
	NBP of naphthalene	217.96	491.11
	NMP of tin	231.85	505.00
	NBP of benzophenone	305.90	579.05
	NMP of cadmium	320.90	594.05
	NMP of lead	327.30	600.45
	NMP of zinc	419.50	692.65

other thermometers in a manner that is described in the next section.

Although the ideal gas temperature scale is independent of the properties of any one particular gas, it still depends on the properties of gases in general. To measure a low temperature, a gas must be used at that low temperature. The lowest ideal gas temperature that can be measured with a gas thermometer is about 1°K, provided low-pressure helium is used. *The temperature T = 0 remains as yet undefined.*

§1-4 The international temperature scale. In everyday life in many parts of the world, the *Celsius* temperature is used. This scale, formerly called the *centigrade scale,* has a degree of the same size as that on the Kelvin scale, but its zero point is shifted so as to make the Celsius temperature of the triple point of water be 0.01°C. Using t to designate the Celsius temperature, it may be calculated from the Kelvin temperature by the relation

$$t = T - 273.15°. \tag{1-4}$$

Celsius temperatures of fixed points are listed in the third column of Table 1-2.

At the meeting of the Seventh General Conference of Weights and Measures in 1927, at which 31 nations were represented, an international temperature scale was adopted, not to replace the Celsius or ideal gas scales, but to provide a scale that could be easily and rapidly used to calibrate scientific and industrial instruments. Slight refinements were incorporated into the scale in a revision adopted in 1948. The international scale agrees with the Celsius scale at the basic fixed points listed in Table 1-2. At temperatures between these fixed points the departure from the Celsius scale is small enough to be neglected in most practical work. The temperature interval from the oxygen point to the gold point is divided into three main parts, as follows:

(1) *From 0 to 660°C.* A platinum resistance thermometer with a platinum wire whose diameter must lie between the limits 0.05 and 0.20 mm is used. The relation between the resistance and the temperature is given by the formula

$$R_t = R_0(1 + At + Bt^2),$$

where the constants R_0, A, and B are determined by measurements at the ice point, steam point, and sulphur point.

(2) *From −190° to 0°C.* The same platinum resistance thermometer is used, and the temperature is given by the formula

$$R_t = R_0[1 + At + Bt^2 + C(t - 100)t^2].$$

R_0, A, and B are the same as before, and C is determined from one measurement at the oxygen point.

(3) *From 660 to 1063°C.* A thermocouple, one wire of which is made of platinum, the other being an alloy of 90 percent platinum and 10 percent rhodium, is maintained with one junction at 0°C. The diameter of each wire must lie between 0.35 and 0.65 mm. The temperature is given by the formula

$$\mathcal{E} = a + bt + ct^2,$$

where *a, b,* and *c* are determined from measurements at the antimony point, silver point, and gold point.

For temperatures higher than the gold point, an optical pyrometer is used. The intensity of the radiation of any convenient wavelength is compared with the intensity of radiation of the same wavelength emitted by a black body at the gold point. The temperature is then calculated with the aid of Planck's radiation law. (See §4-2.)

In North America and in Great Britain, the Fahrenheit scale is used in everyday life and the related Rankine scale is used in engineering. These scales are now defined in terms of the Kelvin scale, as shown in Fig. 1-9.

§1-5 Heat, a form of energy. We have seen that the temperature of a system is the property of the system that determines whether it will be in thermal equilibrium with any other system with which it is put in thermal contact. Suppose that system *A,* at a higher temperature than system *B,* is put in contact with *B.* When thermal equilibrium has been reached, *A* is found to have undergone a temperature decrease and *B* a temperature increase. It was therefore quite natural for the early investigators in this field to assume that *A* lost something and that this "something" flowed into *B.* Nowadays we recognize that there is a *flow of energy* from the hotter to the colder body, and we call this energy, *while in transit, heat.*

Heat is the energy transfer between two systems that is con-

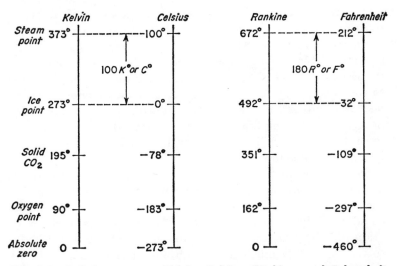

FIG. 1-9 Relations among Kelvin, Celsius, Rankine, and Fahrenheit temperature scales. Temperatures have been rounded off to the nearest degree.

nected *exclusively with the* temperature difference *between the systems.*

It follows therefore that "temperature" and "heat" mean two entirely different things. Temperature is a property of matter, and heat is energy that is flowing because of a temperature difference. These two words are often misused in newspapers and everyday speech. On a hot day, one often reads that there was a "heat" of 90°F. A possible reason for this error lies in the fact that the verb "to heat" is often used to mean "to raise the temperature of": to "heat" water is to raise its temperature. It is best to avoid the use of "heat" as a verb and to use it only as a noun to represent energy in transit. It is always possible to check to see whether one is using the word "heat" properly. If it goes with such words as flow, transfer, enter, leave, absorb, or reject, it is used correctly; otherwise it is not.

When heat enters a system, the internal energy of the system is increased. The transfer of heat, however, is not the only method of changing the energy of a system. When work is done on a system, its energy is also increased. Heat and work therefore constitute two different methods of adding energy to or extracting energy from a system. After the transfer of energy has occurred, it is impossible to say whether the energy that now resides within the system is the result of heat or work. All one can do is to refer to the "energy" inside the system. It is clear, therefore, that *no meaning* can be attached to the unfortunate expressions "the heat in a body" or "the work in a body." It is impossible to separate the energy inside a body into two parts, one due to heat and one due to work.

Suppose you were vacationing at a lake in the mountains and a heavy rainstorm occurred. If someone were to ask at the end of the storm, "How much rain is there in the lake?" no answer could possibly be given. For once the rain ceases, there is no "rain" in the lake, only water. How much of the water is due to rain, and how much to underground springs, cannot be determined. So it is with heat.

§1-6 Isothermal and adiabatic processes. If a thermometer, in thermal equilibrium with a system, gives a constant reading while the system changes, the process is said to be *isothermal*. In Fig. 1-10(a) an isothermal compression of a gas is shown on a graph in which the *pressure P* is plotted against the *volume V*. As the pressure is increased from 1 to 2, the volume is decreased. The graph (b) shows the same process on a *P-T* diagram. Anyone who has pumped up the air in a bicycle tire realizes that the process is usually far from isothermal. To keep constant the temperature of a gas while it is being compressed, it is necessary to allow heat to escape.

A very important type of substance about which we shall have much to say is a *paramagnetic salt*. Most paramagentic salts used by low-temperature physicists are double sulfates with a "magnetic ion" completely surrounded by a lot of water molecules and other molecules that are nonmagnetic. The magnetic ions play the role of little magnets that may be partially oriented by action of an external *magnetic field* of strength H. The degree of orientation of these magnets is measured by a quantity called the *magnetization M*. When the temperature is kept constant during the magnetizing process, the "isothermal magnetization" may be represented on an *H-M* diagram, as is shown in Fig. 1-10(c), from 1 to 2. As in all isothermal processes, heat must be allowed to flow, and the heat that *leaves* during an isothermal magnetization is indicated in graph (d), where the process is drawn on an *H-T* graph.

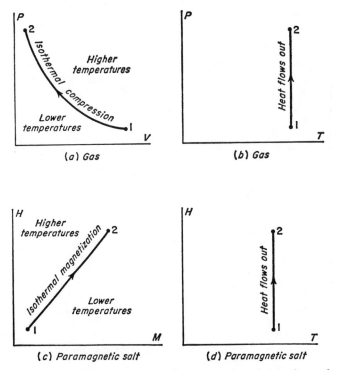

FIG. 1-10   (a) and (b) Two ways of representing the isothermal compression of a gas. (c) and (d) Two ways of representing the isothermal magnetization of a paramagnetic salt.

Isothermal processes, as a rule, must be carried out rather slowly, to allow heat either to leave or to enter. Without a suitable heat flow, the temperature would not remain constant. When a process is performed rapidly, or when a system undergoes a process while it is surrounded by an adiabatic wall, heat cannot enter or leave, and the process is said to be *adiabatic*. The temperature of a system changes during an adiabatic process. When a gas is compressed adiabatically, work is done *on* the gas, thereby increasing the internal energy and the temperature. During an adiabatic expansion, however, the gas does work, loses energy, and undergoes a drop in temperature. In Fig. 1-11(a) an

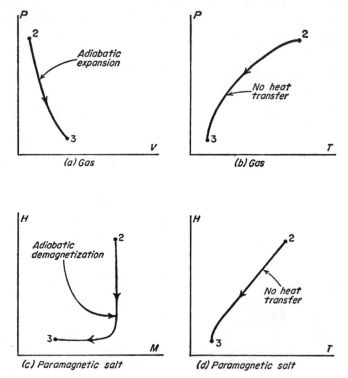

FIG. 1-11   (a) and (b) Two ways of representing the adiabatic expansion of a gas. (c) and (d) Two ways of representing the adiabatic demagnetization of a paramagnetic salt.

adiabatic expansion of a gas is shown on a P-V diagram. Comparing the curve with that in Fig. 1-10(a), it is seen that an adiabatic curve is steeper than an isothermal one. The same adiabatic expansion of Fig. 1-11(a) is shown on a P-T diagram in graph (b), where the temperature drop associated with adiabatic expansion is emphasized.

An adiabatic *demagnetization* of a paramagnetic salt is depicted on an *H-M* diagram in Fig. 1-11(c). This same process drawn on graph (d) shows the temperature drop similar to that produced by the adiabatic expansion of a gas.

§1-7 **The Kelvin temperature scale.** The nice thing about isothermal and adiabatic processes is that they may be accomplished by *all* systems, electric capacitors, electric cells, liquids and solids, even surface films and wires, as well as gases and paramagnetic salts. All that is required of a system during an isothermal process is that heat be allowed to enter or leave so as to keep the temperature constant. All that is required of a system during an adiabatic process is that *no* heat be allowed to enter or leave.

The various values of pressure *P* and temperature *T* assumed by a gas when it undergoes an isothermal or an adiabatic process may be measured accurately, so that the graphs shown in Figs. 1-10 and 1-11 could have been drawn with accurate numerical values of *P* and *T* indicated on the axes. Consider the pressure-temperature graph shown in Fig. 1-12, where the ideal gas scale is indicated temporarily by the symbol θ. Suppose an experi-

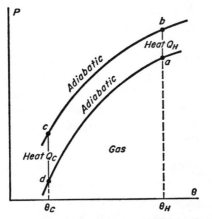

FIG. 1-12 The lines *ab* and *cd* represent isothermal processes that lie between the *same* two adiabatics. The heat Q_H is transferred at the "hot" temperature, whereas Q_C is transferred at the "cold" temperature. The ratio of the Kelvin temperatures T_H/T_C is *defined* to be equal to the ratio of the heats Q_H/Q_C.

menter set the temperature and pressure of a constant amount of gas at the values corresponding to the point marked *a*. By allowing the gas to expand slowly and adiabatically, he could measure many different pressures and corresponding temperatures on his way down to point *d*, thereby enabling him to draw an accurate, smooth curve *ad*. If he set the original pressure and temperature of his gas at the point *b* and then allowed his gas to expand slowly and adiabatically, he could locate enough P-θ points so as to draw another smooth, accurate adiabatic curve *bc*. Let us therefore regard the two adiabatics in Fig. 1-12 as two accurately determined curves among many that could be obtained.

Now let us imagine an accurate measurement of the *heat* that must enter or leave the gas in order to proceed *isothermally* from a point on one adiabatic, such as *a*, to the corresponding point on the other adiabatic, *b*. Denote this amount of heat by the symbol Q_H. It is a matter of convention to choose Q_H positive to represent heat that enters. Let us do the same thing for another isothermal process such as *cd between the same two adiabatics* and call the heat Q_C. (The subscripts H and C refer to "hot" and "cold," relatively speaking.)

The same set of measurements needed to locate two adiabatics and to measure two heats, Q_H and Q_C, transferred at two given temperatures between the same two adiabatics, can be made on *any system whatever*. What Kelvin did was to show, with the aid of the second law of thermodynamics, that the *ratio* Q_H/Q_C *is the same for all systems* which undergo isothermal processes at the same two temperatures, provided of course that the two isothermal processes of any one system lie between the same two adiabatics. Thus:

$$\frac{Q_H}{Q_C} = \frac{\text{Some function of the "hot" temperature}}{\text{The } same \text{ function of the "cold" temperature}}.$$

This ratio was chosen by Kelvin to be the ratio of what is known now as the *Kelvin temperatures*, T_H/T_C, or

$$\frac{T_H}{T_C} = \frac{Q_H \text{ (isothermal process between two adiabatics)}}{Q_C \text{ (isothermal process between the same two adiabatics)}}.$$

It should be emphasized that T_H/T_C was arbitrarily chosen to be the first power of the ratio Q_H/Q_C. An equally legitimate temperature scale could have been defined by choosing any single-valued function of Q_H/Q_C.

When one of the isothermal processes is performed at the temperature of the triple point of water and this temperature is arbitrarily chosen to be $T_3 = 273.16°K$, as in the case of the ideal gas scale, the Kelvin temperature T associated with any other isothermal process is given by

$$T = 273.16° \frac{Q}{Q_3},$$ (1-5)

which should be compared with the definition of the ideal gas scale of temperature:

$$273.16° \lim_{P_3 \to 0} \frac{P}{P_3}.$$

Kelvin was able to give a simple proof that $Q/Q_3 = \lim(P/P_3)$. It follows that the *ideal gas temperature scale, over the range of pressure and temperature within which a gas thermometer can be used, is identical with the Kelvin scale.* This is the reason we use the name of Kelvin and the one symbol T for a temperature on the ideal gas scale.

It follows from Eq. 1-5 that the heat transferred isothermally between two given adiabatics decreases as the temperature decreases. Conversely, the smaller the value of Q, the lower the corresponding T. The smallest possible value of Q is zero, and the corresponding T is absolute zero. *Thus, if a system undergoes a reversible isothermal process without transfer of heat, the temperature at which this process takes place is called* **absolute zero.** In other words, at absolute zero, an isotherm and an adiabatic are identical.

The definition of absolute zero holds for all substances and is therefore independent of the peculiar properties of any one arbitrarily chosen substance. Furthermore, the definition is in terms of purely "large-scale" concepts. No reference is made to molecules or to molecular energy. The statement that is found so often in elementary books and newspaper articles, that at the temperature $T = 0$ all molecular activity ceases, is entirely erroneous. Modern atomic theory shows that the atoms of a solid (or a liquid in the case of helium) at absolute zero have a store of kinetic energy, called *zero point energy,* which may be considerable. As a matter of fact, the zero point energy of liquid helium is so large (three times as large as the heat of vaporization) that a crystal of helium under its own vapor pressure would be unstable. Under pressure, however, the reduction in volume brings the helium atoms nearer together so that the fields of force may interlock and a crystalline solid may form.

The temperature $T = 0$, absolute zero, is the lowest and coldest imaginable. There is no sense whatever in the conception "colder than absolute zero."

2. Temperature, Entropy, and Disorder

§2-1 Energy and entropy. We have been using the term "energy" as though it were a completely familiar concept to everyone. All men think they are good automobile drivers, and all women think they are good cooks. Similarly, people who have been using the word "energy" all their lives are sure they know its meaning. In a certain sense they do. Familiarity may breed contempt, but it also breeds some understanding. Without being clear about mathematical operations, the average person senses that a system has a large amount of energy (1) if it took a lot of work to put the system in its present state, or (2) if it is capable, in its present state, of performing a large amount of work. These ideas are correct and require, for the purposes of our discussion, only a small amount of amplification.

The internal energy U of a system has the limitation that *only its change* can be calculated. When we write the energy U alone, we automatically mean the difference $U - 0$, that is the difference between the energy in the present state and that in *some arbitrarily chosen* state from which the energy was reckoned. Suppose a system undergoes a process in which the changes in its properties are very small, such as a small change of volume dV, a small change of temperature dT, etc. If a small amount of work dW is done *by* the system *adiabatically*, the energy change dU must necessarily be equal to $-dW$. Thus

$$dU = -dW \text{ (adiabatic)}$$

is a real operational definition, as all good definitions in physics must be. It tells us that the system should be surrounded by an adiabatic wall and that work should be done. The amount of work done *on* the system $-dW$ is then equal to the internal energy change. When heat dQ is allowed to *enter* at the same time, then dU will consist of two parts, one due to the work and the other to the heat, or

$$dU = dQ - dW. \qquad (2\text{-}1)$$

Eq. 2-1 expresses what is called the *first law of thermodynamics*. It is apparent that, if no heat is transferred and no work is done, $dU = 0$, or the energy remains constant, as it should, for the system is then isolated. The conventional choice of signs, dQ positive for heat put in and dW positive for work put out, is due to the fáct that the subject developed originally for the theory of heat engines.

The change in internal energy may also be regarded as the heat transferred, but *not in general*—only for a process in which no work is done. It is therefore an interesting and important fact that nature provides us with another quantity called *entropy* (not as familiar as energy, but just as easy to understand) whose change is connected with the heat transferred whether work is done at the same time or not. There is only one hitch, however, and that is that the process must be reversible.

A *reversible process* is one in which the system and its surroundings are nicely behaved: there are no complications like acceleration, waves, eddies, turbulence, friction, etc. Every property like pressure, temperature, magnetization, etc., is *uniform throughout the system* so that *one* value of a property holds for the entire system. The system is very close to equilibrium at all times. A reversible process, although an idealization that is not perfectly realizable, may be approximated well by doing things slowly.

The fundamental property of the entropy S of a system is that, *in a small reversible process, the product of the temperature and the entropy change is equal to the heat transferred*. In symbols,

$$T\, dS = dQ. \qquad (2\text{-}2)$$

Eq. 2-2 may be taken to be a statement of *the second law of thermodynamics*. Evidently, when a process is both reversible and adiabatic, then $dS = 0$ and S is constant. *A reversible adiabatic process takes place at constant entropy and is called isentropic.*

The reader is asked to accept as plausible the existence of a function S with the fundamental property given by Eq. 2-2. If he is to agree, he should at least be told how one goes about calculating the change in this function and, if this calculation is made, what sort of quantity is obtained. The answer is contained in Eq. 2-2 for, if it is rewritten in the form

$$dS = \frac{dQ}{T}, \tag{2-3}$$

we realize that, if we can express the heat dQ that is transferred in a reversible process in terms of T, then the equation can be integrated between the arbitrarily chosen standard state (at which S is given the arbitrary value zero) and any other state.

The expression for dQ is quite complicated for some systems, but for an ideal gas it has a particularly simple form, namely,

$$dQ = C_v \, dT + P \, dV,$$

where C_v represents the molar heat capacity at constant volume. If this expression for dQ is substituted into Eq. 2-3, we get, for n moles,

$$dS = nC_v \frac{dT}{T} + \frac{P}{T} \, dV.$$

But, for an ideal gas, $PV = nRT$ where R is the universal gas constant. Hence,

$$dS = n \left[C_v \frac{dT}{T} + R \frac{dV}{V} \right].$$

Integrating from an arbitrarily chosen standard state $T = T_0$, $V = V_0$ at which we set $S = 0$, to any other state (T,V,S), we get

$$S = n \left[C_v \ln \frac{T}{T_0} + R \ln \frac{V}{V_0} \right], \tag{2-4}$$

where T_0 and V_0 are constant. Although Eq. 2-4 holds only for an ideal gas, it has characteristics very similar to those of the entropies of practically all solids, liquids, and gases, namely:

(1) If only the temperature is varied, the higher the temperature, the greater the entropy.

(2) If only the volume is varied, the larger the volume, the greater the entropy.

The implications of these results are important. Consider the *isentropic expansion* of a gas. An increase of volume provides a positive contribution to an entropy change. To keep the entropy constant, there must be a compensating negative contribution to the entropy change, which is provided by a *decrease in temperature*. The drop in temperature that takes place when a gas expands adiabatically may therefore be regarded as an effect that is needed to offset the effect of the rise in volume, in order to keep the entropy constant.

When we combine the first and second laws of thermodynamics, Eqs. 2-1 and 2-2, we get

$$dU = T \, dS - dW. \tag{2-5}$$

Under conditions in which *no work* is done,

$$\boxed{T = \frac{dU}{dS} \text{ (no work).}} \tag{2-6}$$

Eq. 2-6 provides us with a method of measuring temperature that is applicable in a temperature range in which no ordinary thermometer will function properly. The ratio of a small energy change to the accompanying small entropy change gives the Kelvin temperature. This is particularly useful in the region of extremely low temperatures, as we shall see later.

Although the relation between temperature, entropy change, and heat given by Eq. 2-2 holds only for a reversible process, the entropy concept plays an equally important role in *irreversible processes,* that is, those in which the system passes through a succession of states which are not capable of being described with the aid of a few, large scale, uniform properties, and where friction, electrical resistance, inelasticity, and other such dissipative processes occur. In all irreversible processes

$$T \, dS > dQ,$$

and therefore, if our system includes everything that is changing or interacting during an irreversible process, there is no heat transfer between this isolated system and its surroundings, so that $dQ = 0$, and

$$dS > 0,$$

which means that *S increases.* As an isolated system undergoes an irreversible process on its way toward equilibrium, its entropy increases. *The equilibrium state is the state of maximum entropy.*

§2-2   **The molecular point of view.** Up to this point our treatment of the behavior of material systems has been in terms of large scale or *macroscopic* properties of matter, such as pressure, volume, temperature, and magnetization. There are, however, distinct advantages in recognizing that matter is composed of molecules and in describing the states of matter in terms of molecular behavior. Every macroscopic property has a molecular interpretation. Thus the pressure exerted by a gas is equal to the rate of change of momentum of its molecules in collisions against unit area of the containing vessel. The magnetization of a paramagnetic solid is related to the degree of orientation of the atomic magnets of the sample when placed in an external magnetic field. By adopting the molecular point of view, we may not only calculate the macroscopic properties of a system from a knowledge of the mechanical, electrical, and magnetic properties of molecules, but we may also obtain a greater insight into the meaning and importance of the concepts of temperature, energy, and entropy. The branch of physics which is concerned with applying probability concepts to interpret thermal behavior of matter is called *statistical mechanics.*

As an illustration of the molecular point of view, let us limit ourselves to a system with the following characteristics:

(1) It consists of an enormous number N of identical particles, in the neighborhood of 10^{20}.

(2) Each molecule is *almost* independent of the others. This means that, if the particles collide, they do so only occasionally, or if they do not collide, they interact *weakly* with one another through mutual electrical or magnetic fields.

(3) Each molecule is capable of existing in only special discrete states whose energies are u_0, u_1, u_2, and so on.

A microscopic description of the over-all *state* of a system of N molecules is provided by specifying that there are:

n_0 molecules in the state 0 with energy u_0,
n_1 molecules in the state 1 with energy u_1,
n_2 molecules in the state 2 with energy u_2,

and so on. The values of the n's are called the *populations* of the states. The populations depend upon how close to equilibrium the system is, and as the system approaches equilibrium, the populations change until they reach their equilibrium values. It is our purpose to understand how to find these equilibrium values.

Suppose we have only three molecules, A, B, and C, each of which may exist in only four states $u_0 = 0$, $u_1 = \epsilon$, $u_2 = 2\epsilon$, and $u_3 = 3\epsilon$. Suppose the total energy of our system of three molecules must remain constant at the value 3ϵ. Fig. 2-1 shows three possible

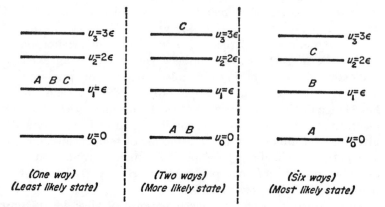

(One way)   (Two ways)   (Six ways)
(Least likely state)   (More likely state)   (Most likely state)

FIG. 2-1   Three different over-all states of a system composed of molecules, A, B, and C, each of which has four energy states 0, ϵ, 2ϵ, 3ϵ. The system has the total energy 3ϵ in each of the three over-all states depicted.

states in which our system may exist. That on the left is characterized by the populations $n_0 = 0$, $n_1 = 3$, $n_2 = 0$, and $n_3 = 0$. This may be achieved in only one way, namely, by having each of the molecules in the state $u_1 = \epsilon$.

The state shown in the middle of Fig. 2-1 with $n_0 = 2$, $n_1 = 0$,

$n_2 = 0$, and $n_3 = 1$ may be achieved in three different ways, since A or B or C could be at the top level.

The state shown at the right of Fig. 2-1 is the most probable of the three, because when any one of the three molecules is in the bottom level, the others may have either of two positions. There are therefore six different ways of achieving this state.

It is a matter of elementary algebra to show that the number of ways of achieving a state, or the *thermodynamic probability of a state* $\mathcal{P}$ is given by

$$\mathcal{P} = \frac{N!}{n_0!\, n_1!\, n_2!\, \ldots}. \qquad (2\text{-}7)$$

Using the states of Fig. 2-1,

$$\text{(left)} \quad \mathcal{P} = \frac{3!}{0!\, 3!\, 0!\, 0!} = 1,$$

$$\text{(middle)} \quad \mathcal{P} = \frac{3!}{2!\, 0!\, 0!\, 1!} = 3,$$

$$\text{(right)} \quad \mathcal{P} = \frac{3!}{1!\, 1!\, 1!\, 0!} = 6.$$

(It will be remembered that $0! = 1$, because there is just one way of having no molecules in a particular state.)

If a system of 10^{20} molecules each of which is capable of existing in an infinite number of energy states is left to itself, subject only to the restrictions of constant number and constant energy, it will end up in the state of highest probability. This is the equilibrium state. But we have already learned that, in an equilibrium-seeking process, the entropy also increases. Evidently there must be some connection between the macroscopic property of entropy and the microscopic property of thermodynamic probability. This relation, which was discovered by Ludwig Boltzmann, is

$$S = k \ln \mathcal{P}, \qquad (2\text{-}8)$$

where k is known as Boltzmann's constant ($k = 1.38 \times 10^{-6}$ erg/degree).

Both the entropy and the thermodynamic probability may be considered from a slightly different point of view, that of *molecular disorder*. If a gas occupies a very small volume, it is spatially set apart; all the molecules are in a restricted region, and no molecules are anywhere else. This is a state of very low *disorder*, like a neat bureau drawer where all the handkerchiefs are on the left and all the socks (or stockings) are on the right. If the gas is allowed to expand so that the molecules occupy a larger space, they get more mixed up, more uniformly distributed throughout space. This is a state of greater disorder, like the bureau drawer after a teen-ager has used it for a few days.

As a system proceeds toward its equilibrium state, it goes through states of greater and greater disorder. There is a tendency in nature to proceed toward a state of greater disorder: rocks weather and crumble, iron rusts, machines wear out, and worst of all, people get less efficient. These are all complicated examples of the fact that the tendency of material systems toward a state of greater disorder is just another way of stating what was said before in terms of increasing entropy and increasing thermodynamic probability. All three of these terms refer to the same aspect of nature.

Let us now return to the mathematical problem of equilibrium: Given a system consisting of an enormous number N of particles which interact with one another only very weakly and which are capable of existing in states of which the energy levels are u_0, u_1, u_2, ... If at any moment the populations of these states are n_0, u_1, u_2, ..., what will the populations be when equilibrium is reached under conditions such that the total number of particles N is constant and the total energy U is constant?

We have to *maximize* the entropy S, which, from Eqs. 2-7 and 2-8, is

$$S = k \ln \frac{N!}{n_0! \, n_1! \, n_2! \dots} = \text{a maximum,}$$

subject to the conditions that

$$n_0 + n_1 + n_2 + \dots = N = \text{constant,}$$

and

$$n_0 u_0 + n_1 u_1 + n_2 u_2 + \dots = U = \text{constant.}$$

This is a simple mathematical problem that is solved by the

"Lagrange method of undetermined multipliers." It is helpful to use Stirling's approximation

$$\ln (x!) = x \ln x - x.$$

Let us skip the details and jump to the answer. At equilibrium the populations are given by $n_0 = Ke^{-\beta u_0}$, $n_1 = Ke^{-\beta u_1}, \ldots$ or, for any typical energy level, say the ith,

$$n_i = Ke^{-\beta u_i}, \tag{2-9}$$

where K and β are constants whose physical significance must be determined.

To eliminate K, we divide by $n_0 = Ke^{-\beta u_0}$ and get

$$\frac{n_i}{n_0} = e^{-\beta(u_i - u_0)}. \tag{2-10}$$

The constant β is very interesting. Its physical significance is discovered by examining the equilibrium of a *composite* system composed of our original system (populations n_0, $n_1 \ldots$, and energy states u_0, $u_1, \ldots$) *and* a second system with populations n'_0, $n'_1, \ldots$, and energy states u'_0, $u'_1, \ldots$. If the two systems are gases, they may be mixed together. If they consist of a crystal and a gas, they must be separated by a diathermic wall. Now the total number of molecules in *each* part of our composite system must remain constant. Thus

$$n_0 + n_1 + \ldots + n_i + \ldots = \text{constant},$$

and

$$n'_0 + n'_1 + \ldots + n'_j + \ldots = \text{constant}.$$

But the energy of each part of the composite system is *not* constant. Instead, the *total energy* of the whole composite system must remain constant, or

$$n_0 u_0 + n_1 u_1 + \ldots + n_i u_i + \ldots$$
$$+ n'_0 u'_0 + n'_1 u'_1 + \ldots + n'_j u'_j + \ldots = \text{constant}.$$

When the problem of equilibrium is now solved, we get two sets of equations like Eq. 2-9, namely,

$$n_i = Ke^{-\beta u_i} \quad \text{and} \quad n'_j = K'e^{-\beta u'_j},$$

where the n values, the u values, and K refer to the first part of the composite system and the n' values, the u' values, and K' refer to the second part. But the striking fact is that both equations have the same β. With a system composed of only one set of molecules, equilibrium is determined by one set of equations with one value of β. With two sets of molecules in equilibrium through a diathermic wall, there are two sets of equations, *but the β in each set is the same!* The conclusion that the quantity β is connected with the temperature is inescapable.

It is a simple matter to derive the relation between β and T. Since, for one system,

$$S = k \ln \frac{N!}{n_0!\, n_1!\, n_2!\, \ldots}$$

and

$$n_i = Ke^{-\beta u_i},$$

the various n's may be substituted into the expression for S and and use may be made of the facts that

$$\sum n_i = N$$

and

$$\sum n_i u_i = U.$$

When all this is done, we get the result that

$$S = Nk \ln \sum e^{-\beta u_i} + k\beta U. \tag{2-11}$$

But it was shown in Eq. 2-6 that when no work is done

$$\frac{1}{T} = \frac{dS}{dU}.$$

Therefore, differentiating Eq. 2-11 with respect to U and setting the derivative equal to $1/T$, we get the beautiful result

$$\beta = \frac{1}{kT}.$$

Introducing this value of β into Eq. 2-10 for the two energy levels i and 0, we get

$$\boxed{\frac{n_i}{n_0} = e^{-\frac{u_i - u_0}{kT}}.} \tag{2-12}$$

This is known as *Boltzmann's equation*, which tells the ratio of the population of *any* energy state to that of the first or lowest energy state. Since $u_i - u_0$ is positive and kT is positive, it follows that n_i is less than n_0, or the population of an upper state is always less than that of any state of lower energy. The higher the energy, the smaller the population—for all positive values of the temperature. If the temperature were infinite, all the levels would be populated equally, but it would require an infinite amount of energy for this purpose, since there are an infinite number of levels.

The total energy U is equal to

$$U = \sum n_i u_i,$$

and, upon substituting the values of the n's given by Boltzmann's equation, we get

$$U = K \sum u_i e^{-u_i/kT}.$$

By making use of our knowledge of energy states and by converting the sum to an integral, it is possible to evaluate the total energy U and to subtract various terms that refer to potential energy. We finally end up with a fairly good expression for the *kinetic energy E_k* of the molecules of a substance. In a rough sort of way, the results are shown in Fig. 2-2. The kinetic energy is proportional to T only when the substance is a gas! This is by

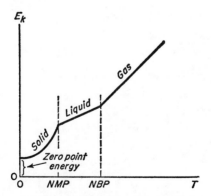

FIG. 2-2   Rough graph showing kinetic energy as a function of temperature, and zero-point energy.

no means the case in the liquid and solid regions, and at absolute zero the kinetic energy is not zero, but a finite amount—the *zero point energy*.

Since molecular kinetic energy and temperature are so closely related (proportional to each other only in the case of an ideal gas), it is perfectly legitimate and, at times, very helpful to think of temperature as a measure of molecular agitation of a random, disorderly kind. If the little magnetic domains of a piece of iron are nicely lined up in the direction of a magnetic field, this represents a state of low entropy or low disorder. To increase the disorder, all that is necessary is to raise the temperature.

One of the greatest advantages of the molecular point of view culminating in Boltzmann's equation (Eq. 2-12) is that it enables us to associate a value of T with any system (fairly near equilibrium, of course) whose molecules are distributed among energy levels. We shall see later that a temperature may be associated in this way with the electrons in a metal, with the electronic velocity distribution in a gaseous electric discharge, and with the distribution of nuclear magnets among their energy levels.

3. The Approach to Absolute Zero

§3-1   The Joule-Kelvin effect. The introductory part of any subject is bound to be difficult. There are always more definitions, more fundamental concepts, and more theorems than one expects. It is a pleasure to announce at this point that we now have sufficient background in terminology and theory to enable us to understand some of the most spectacular and fruitful researches in modern physics, namely, the investigations of the very cold and of the very hot. Let's go down first.

To obtain a moderate amount of cooling, say from 300°K (room temperature) to 255°K (−18°C, or 0°F), the most effective method would be to cause a gas, contained in a cylinder equipped with a movable piston, to undergo an adiabatic, almost reversible expansion. Since work is done at the expense of the internal energy of the gas, the temperature drops. This method has two disadvantages:

1) It requires pistons moving in cylinders, and therefore presents problems of lubrication, vibration, and noise;

(2) As the gas gets colder, the temperature drop for a given pressure drop is smaller.

It is therefore of great significance that a temperature drop may be produced by a process which involves no moving pistons and which gets better as the temperature goes down. The process is one in which a fluid (gas or liquid) at high pressure seeps through a tiny opening or series of openings, *adiabatically and irreversibly,* into a region of lower pressure. This is called a *throttling process,* and the temperature change accompanying a throttling process is known as the *Joule-Kelvin* (or Joule-Thomson) *effect.* In some older books the process is called the *porous plug experiment.* It is the cooling process that is used in every

37

home refrigerator, electric or gas, and has therefore been the subject of much engineering research and development.

If a liquid about to vaporize undergoes a throttling process, a cooling effect accompanied by partial vaporization always occurs. With a gas, however, cooling occurs only if the initial pressure and temperature of the gas (indicated by a point on a T-P diagram) lie *inside* a curve called the *inversion curve,* shown for hydrogen in Fig. 3-1. Thus, when hydrogen gas at a pressure of

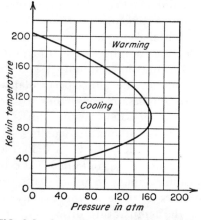

FIG. 3-1   Inversion curve for hydrogen.

100 atmospheres undergoes throttling to any lower pressure, cooling results only when the original temperature is equal to or lower than 160°K. But 160°K is much lower than room temperature. Therefore, to cool hydrogen from room temperature to its liquefying point, 20°K, we must first either use the reversible adiabatic expansion method to get below 160°K or employ a coolant such as liquid nitrogen. Then the throttling process may be used to give further cooling. It is, however, still a long step from 160°K to the liquefaction point, 20°K—too long a step to be accomplished by one throttling process. Repeated use of the Joule-Kelvin effect is then used to liquefy a gas.

A typical liquefaction machine is shown symbolically in Fig. 3-2. The gas to be liquefied is compressed and then sent through a coil of pipe immersed in a cooling liquid. The coolant must be cold enough to lower the temperature of the gas to a value within the inversion curve. Ordinary tap water is sufficient for nitrogen gas, but liquid nitrogen (70°K) must be used to cool hydrogen gas, and liquid hydrogen (20°K) to cool helium gas. When the gas leaves the cooler, it enters one section of a double pipe, called a *heat exchanger*. At the other end of the heat exchanger it

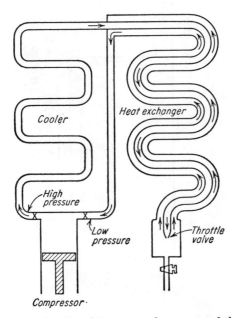

FIG. 3-2   Apparatus for liquefying a gas by means of the Joule-Kelvin effect.

emerges through a narrow opening (*throttling valve*) and undergoes throttling with an accompanying temperature drop—not enough for liquefaction. The cooled gas now passes in the opposite direction through the outer section of the heat exchanger and acts to cool off the incoming stream of gas that has not yet reached the throttling valve. When this cooled gas undergoes throttling, it starts at a lower temperature; its temperature drop is therefore larger, and it emerges still colder and serves to cool

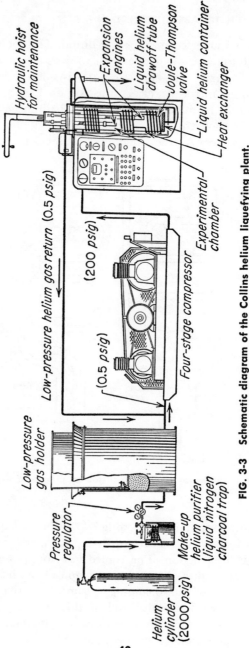

FIG. 3-3 Schematic diagram of the Collins helium liquefying plant.

the incoming stream even more. After a while a steady temperature difference is set up in the heat-exchanger: the coolant temperature at one end and the liquefaction temperature at the throttle end. The liquid collects in the well and is drawn off at convenient intervals.

In the "old days" of cryogenics (low-temperature physics) physicists had to make their own liquid helium. The ordeal usually started early in the morning when liquid nitrogen was used to cool off the hydrogen liquefier. By about noon, enough liquid hydrogen had collected to allow the helium liquefier to be cooled off. If there were no explosions and no pipes clogged with solid air, liquid helium started to collect at about 6 P.M., and by about eight in the evening the experiment was ready to start. If everything went well, the experimenters got to bed by three or four the next morning.

The situation today is quite different. After a few preliminary operations that can be performed by a technician, a machine designed by Samuel Collins provides liquid helium without the necessity of liquefying hydrogen first and even without liquid nitrogen, although it is a bit faster if liquid nitrogen is used. In the Collins machine, the helium is lowered to a temperature within the inversion curve by expanding adiabatically and almost reversibly against a piston in a cylinder. The remainder of the cooling and liquefaction are then accomplished with the aid of the Joule-Kelvin effect. All the apparatus necessary may be bought as a package, put on a truck, delivered, installed, and tested within a period of a week or so. A schematic diagram of a Collins helium liquefying plant is shown in Fig. 3-3.

§3-2 **In the liquid helium region.** When a gas is liquefied, the normal boiling point (or, in the case of carbon dioxide, the normal sublimation point) and the triple point temperatures and pressures are usually determined as accurately as possible. Temperatures are measured with a constant-volume helium gas thermometer according to the principles developed in Chap. 1. Once these temperatures and pressures are measured, they serve as convenient fixed points for other thermometers that are simpler to operate. Some of the most important fixed points of low-temperature physics are listed in Table 3-1.

The lowest temperature that can be reached easily with liquid helium is about 1°K. This is achieved by pumping the vapor away as fast as possible through as wide a tube as possible. The remaining liquid and vapor may be regarded as undergoing an approximately reversible adiabatic expansion in which cooling always occurs. With special high-speed pumps, temperatures as low as 0.7°K have been reached, but this is rare. Temperatures lower than 0.7°K cannot be reached by pumping liquid helium because a film of liquid helium creeps up the walls of the pumping tube, extracts heat from these walls, vaporizes, and then recondenses at another place, giving up its heat of condensation.

TABLE 3-1 *Useful Fixed Points in Low-Temperature Physics*

	Normal boiling point (760 mm)	Triple point
Carbon dioxide	(Sublimes) 194.7°K	216.6°K 5.11 atm
Nitrogen	77.32°K	63.14°K 94 mm
Hydrogen	20.37°K	13.96°K 54.1 mm
Helium 4 (ordinary helium)	4.216°K	2.186°K 37.80 mm
Helium 3 (light isotope)	3.195°K	? ?

The effect of all this is the equivalent of a "heat leak," which prevents further cooling.

If the exceedingly rare and expensive light isotope of helium, He³, present in ordinary helium to only a concentration of one part in a million, is obtained pure and is liquified, a temperature as low as 0.3°K can be obtained by reducing its vapor pressure with fast pumps.

It is important to know the vapor pressure of liquid hydrogen and of liquid helium accurately over the entire range from the temperature at which the vapor pressure is too low to measure up to (and even beyond) the normal boiling point, because, when this information is at hand, a measurement of the vapor pressure tells the experimenter what the temperature is. The vapor itself is used as its own thermometer. Thus a hydrogen vapor-pressure thermometer may be used in the range from 20 to 10°K, and a helium vapor-pressure thermometer from 4 to 1°K. The region from 10 to 4°K is more difficult. Table 3-1 shows that there are no fixed points in this region. Fortunately, Plumb at the National Bureau of Standards in 1962 developed a thermometer whose thermometric property is the velocity of sound in helium gas, and which is capable of providing accurate values of Kelvin temperature in the range 10 to 4°K.

The gas thermometer, the vapor-pressure thermometer, and the sound-velocity meter are elaborate, exacting, and sluggish devices. To measure heat capacities, thermal conductivities, and several other physical quantities at low temperatures, many measurements of small temperature changes must be made quickly and accurately. For these purposes secondary thermometers must be used.

One of the first to be employed was a resistance thermometer made of carbon. Pieces of paper with carbon deposited on them or strips of carbon prepared by painting with colloidal suspensions have two advantages. They have extremely small heat capacities and can therefore follow temperature changes quickly, and their electric resistance, which increases rapidly as the temperature is reduced, is insensitive to the presence of a magnetic field. Their main disadvantage is that they must be calibrated anew each time they are used.

In 1951, Clement and Quinell discovered that carbon composition radio resistors, made by Allen-Bradley and rated from $\frac{1}{2}$ to 1 watt, have all the properties most desired in a low-temperature secondary thermometer, namely, high sensitivity, reproducibility, and insensitivity to magnetic fields. The reasons for these desirable properties are not understood, but such thermometers contributed to the accuracy of much of the work done in low-temperature physics in the decade from 1951 to 1961.

Nowadays, semiconductors—for example, germanium adulterated with minute quantities of other substances such as arsenic —are replacing radio resistors as thermometers in the low-temperature range. Semiconductors are quite sensitive and accurately reproducible. Such thermometers have the advantage that the processes responsible for the resistance change are at least partly understood. Magnetic thermometers are also on the market, but more about these later.

One of the most interesting substances in the liquid helium range is liquid helium itself. When most liquids in equilibrium with their vapor are cooled, eventually a temperature and vapor pressure are reached at which a solid begins to form. This is the triple point—that is, for every known substance except helium. When helium and its vapor are cooled to $2.2°K$, known as the *lambda point,* a third phase forms, but it is another liquid, not a solid! The new liquid is designated by Roman numeral II, and helium remains in this new phase all the way from $2.2°K$ to absolute zero! In this range, liquid helium II exhibits the remarkable property known as *superfluidity,* which shows itself in the following four phenomena:

(1) It flows *with practically no friction* through small openings, closely-packed powders, and narrow capillaries, like water through a sieve. If the rate of flow of liquid helium through a capillary is plotted against the temperature, as in Fig. 3-4, the sudden rise at the λ-point is most spectacular.

(2) when a restricted region of liquid helium II undergoes a small rise of temperature due to an embedded heating coil or to the absorption of radiation from a flashlight, a rise in pressure takes place and a fountain of liquid helium spurts up. This phenomenon, known as the *fountain effect,* is shown schematically in Fig. 3-5.

(3) When rapid local variations in temperature take place in a small region of liquid helium II, these variations are propagated as a wave and give rise to temperature variations in a small resistor used as a receiver of the wave. Such thermal waves are called *second sound* and exhibit the phenomena of reflection, diffraction, radiation pressure, etc.

(4) As shown in Fig. 3-6 (a), a small vessel, originally empty, fills up when partially immersed in liquid helium II, whereas the full vessel in Fig. 3-6 (b), when lifted up, empties itself by means of a *film creeping* along the walls of the vessel. When the originally full vessel is removed completely from the liquid helium II,

the creeping film creates drops at the bottom, as shown in Fig.
3-6 (c).

In contrast, the light isotope of helium, He³, has none of these
peculiar properties.

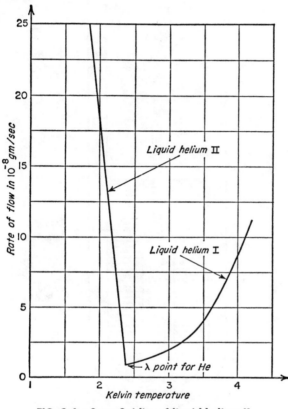

FIG. 3-4   Superfluidity of liquid helium II.

A word about Figs. 3-5 and 3-6 is needed at this point. They
show only the liquid helium and its immediate surroundings, and
none of the tubes, valves, wires, gauges, manifolds, tanks, pumps,
and headaches. The reader is asked to imagine that the helium
dewar is surrounded by liquid nitrogen contained in a larger
dewar that is about four feet long and attached at the top to a

plate through which pumping lines, pressure gauge connections, liquid inlet tubes, gas outlet tubes, electric connecting wires, etc., pass on their way to a bewildering assortment of gadgets. In other

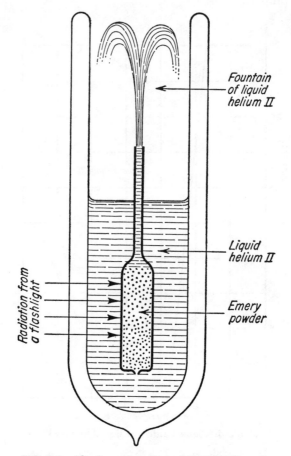

FIG. 3-5   The fountain effect of liquid helium II.

words, the experiments just described are not as easy as they sound or as simple as they look.

§3-3 **Paramagnetic salts.** If it were not for the wonderful magnetic properties of some crystals, it would be impossible to go much lower than the temperatures attainable by pumping on liquid helium. To understand these properties it must first be realized that a component of a crystal lattice may be an ion, that is, an atom that has either lost or gained one or more electrons. Thus, in a crystal of sodium chloride, sodium is present as a positive ion, chlorine as a negative ion. The sodium ion is a sodium atom that has lost its one valence electron and therefore has an electron configuration like an inert gas with completely closed electron shells. Such an ion is not a magnet because all electron spins are compensated.

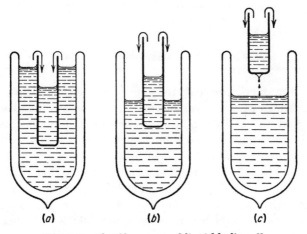

FIG. 3-6 The film creep of liquid helium II.

There are, however, elements which, when in a crystal lattice, form ions whose outer shells are only partially filled. The *uncompensated electron spins* give rise to a magnetic effect, that is, the ion behaves as though it were a tiny magnet. Gadolinium, iron, chromium, and cerium are examples. When these elements are chemically combined with a large number of nonmagnetic molecules or radicals, the individual magnetic ions are so far removed from other magnetic ions that they behave almost independently, like the atoms of an ideal gas, even though they vibrate about equilibrium positions in the lattice. The compounds shown in Table 3-2 are examples of salts each of which has only one magnetic ion surrounded by many nonmagnetic atoms and is therefore, magnetically, very dilute.

TABLE 3-2 *Paramagnetic Salts*

Ion	Compound	Gram-ionic weight (gm)	Curie const. $\left(\dfrac{cm^3 \cdot deg}{gm \cdot ion}\right)$
Gd^{+++}	Gadolinium sulfate		
	Gd$_2$(SO$_4$)$_3 \cdot$8H$_2$O	373	7.85
Fe^{+++}	Iron ammonium alum		
	Fe$_2$(SO$_4$)$_3 \cdot$(NH$_4$)$_2$SO$_4 \cdot$24H$_2$O	482	4.35
Cr^{+++}	Chromium potassium alum		
	Cr$_2$(SO$_4$)$_3 \cdot$K$_2$SO$_4 \cdot$24H$_2$O	499	1.86
Ce^{+++}	Cerium magnesium nitrate		0.318 ($\perp$)
	2Ce(NO$_3$)$_3 \cdot$3Mg(NO$_3$)$_2 \cdot$24H$_2$O	765	0.0056 ($\parallel$)

Quantum mechanics tells the energy levels of magnetic ions, so the equations of statistical mechanics can be evaluated and the magnetization* of a crystal can be computed. This was first done by Brillouin, and the excellent agreement between Brillouin's theoretical equation and experimental results of W. E. Henry is shown in Fig. 3-7. The magnetization is a function of the *ratio* H/T so that the magnetic field needed to produce a moderate magnetization at 3°K is only one one-hundredth that for the same magnetization at room temperature, 300°K. The beginning part of each magnetization curve is seen to be linear so that, for small values of H/T, Brillouin's equation reduces to

$$M = C\frac{H}{T}, \qquad (3\text{-}1)$$

a relation first discovered by Pierre Curie and known as Curie's law. The constant C is called the *Curie constant* and is listed in Table 3-2 along with the gram-ionic weight, which is equal to one-half the molecular weight. The larger the Curie constant the

---

* A compass needle of length l has a magnetic pole of strength m at each end. Its magnetic moment is defined as the product ml. The magnetization of *any* magnetic material of *any* shape is defined as its total magnetic moment.

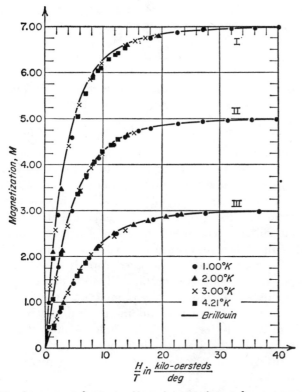

FIG. 3-7 Agreement between Henry's experimental measurements of magnetization and Brillouin's theoretical equation, for (I) gadolinium sulfate, (II) iron ammonium alum, and (III) chromium potassium alum.

smaller the value of H/T that is needed to produce a given magnetization.

For large values of H/T, the curves of Fig. 3-7 bend over, and in the neighborhood of 40,000 oersteds per degree the curves are practically flat, indicating that further increase in H/T produces no appreciable increase in magnetization. When this is the case, magnetic *saturation* is said to take place.

There are two commonly used units of magnetic intensity H, the ampere per meter and the oersted. At the center of a *solenoid* of length l wound with N turns of wire and carrying a current I in a vacuum, the magnetic intensity is

$$H = \frac{NI}{l}.$$

When I is measured in amperes and l in meters, the unit of H is 1 amp/m. The oersted may be defined by the relation

$$1 \text{ amp/m} = 0.0126 \text{ Oe}, \quad 1 \text{ Oe} = 79.4 \text{ amp/m}.$$

§3-4 **Adiabatic demagnetization.** In 1926 it occurred simultaneously to W. F. Giauque in America and to P. Debye in Germany that temperatures below 1°K could be obtained by making use of the magnetic properties of paramagnetic salts. Giauque and Mac-Dougall were the first to perform the experiment. They chose gadolinium sulfate because of its large Curie constant, and mounted it in a glass vessel into which helium gas could be admitted, and which could be evacuated rapidly when necessary. Surrounding this vessel there was liquid helium at a temperature of approximately 1°K, obtained by pumping on the helium and lowering its vapor pressure. Surrounding the liquid helium dewar was liquid nitrogen in its dewar. The whole assembly was narrow enough to fit either between the iron pole pieces of an electromagnet or inside an air-core solenoid. The arrangement is shown symbolically in Fig. 3-8. The paramagnetic salt is usually a conglomeration of crystals packed tightly together and shaped in the form of a sphere or an ellipsoid to achieve favorable magnetic properties.

The experiment is performed by first admitting helium gas to the space containing the paramagnetic salt (the experimental space) to provide heat transfer between the salt and the liquid helium. The current in the magnet coils is then slowly increased, or the magnet is wheeled into position, or the complex of dewars (called the cryostat) is moved between the pole pieces. Any one of these methods provides an *isothermal magnetization* during which heat is transferred from the salt to the liquid helium. This is shown as the process 1 to 2 in Fig. 3-9(b), which is a repetition of Fig. 1-10(d). The parallel between the behavior of a paramagnetic salt and a gas is shown by the neighboring graph, Fig. 3-9(a).

The only difference is that the magnetic field starts at zero, whereas the pressure cannot.

At point 2 the salt is magnetized as much as possible, and it

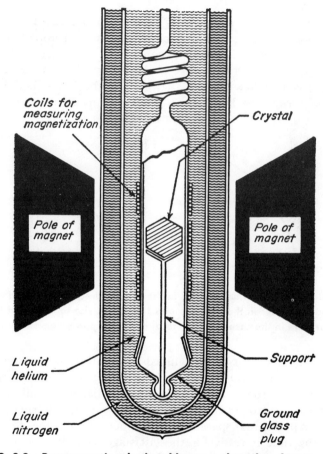

Coils for measuring magnetization

Crystal

Pole of magnet

Pole of magnet

Liquid helium

Support

Liquid nitrogen

Ground glass plug

FIG. 3-8   Paramagnetic salt placed between the poles of a magnet.

is as cold as liquid helium can make it. During the isothermal magnetization process 1 to 2, the heat transferred to the helium causes some of the liquid to boil away. If it were possible to measure this amount accurately, then, knowing the heat of vaporization of liquid helium (about 5 cal/gm), one could find out how much heat left the sample, and since

$$\Delta S = \frac{Q}{T},$$

one could calculate the entropy decrease of the salt. Usually, it is a more accurate procedure to calculate this entropy decrease with the aid of one of Brillouin's equations.

Now comes the crucial part of the experiment—breaking the thermal connection between the salt and the liquid helium by pumping away the helium gas that served as a conducting me-

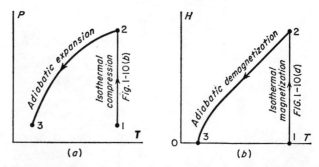

FIG. 3-9 An isothermal process in the proper direction, followed by an adiabatic process in the opposite direction, gives rise to a drop in temperature in the case (a) of a gas and (b) of a paramagnetic salt.

dium. The salt must be surrounded by a really high vacuum, and in a time during which the sample will not warm up appreciably from the heat conducted along wires or radiated by walls. If this pumping out process is accomplished without discovering a leak in the experimental space that was undetected at room temperature, the time is ready for the last stage, namely *adiabatic demagnetization*. This is accomplished either by rolling the magnet away or by swinging the cryostat out of the field. This is process 2 to 3 in Fig. 3-9(b), giving rise to a temperature drop. Plate I is a spectacular view of the equipment. (The narrow portion of the cryostat is about a foot long.)

The processes just mentioned may be plotted on an entropy-temperature graph, as shown in Fig. 3-10. The isothermal magnetization 1 to 2 is accompanied by an entropy *decrease* because heat *leaves* the salt. In the words of the late Professor Simon of Oxford, the isothermal magnetization is an "entropy squeezing" operation, like compressing a gas. The reversible adiabatic de-

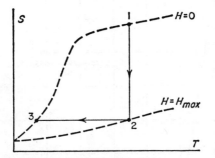

FIG. 3-10 During the isothermal magnetization from 1 to 2, entropy is squeezed out of the salt. During the adiabatic reversible demagnetization from 2 to 3, the entropy remains constant.

magnetization, 2 to 3, like all reversible adiabatic processes, takes place at constant entropy. Fig. 3-11 shows that the lower the temperature of point 1 the lower the temperature of point 3.

It is instructive to analyze the reversible adiabatic demagnetization process from the standpoint of disorder. For the entropy to remain constant, a "disorder-increasing" process like a demagnetization, in which ionic magnets are *disoriented,* must be compensated by a "disorder-decreasing" process, which can be only a drop in temperature.

The degree of orientation of ionic magnets, expressed macroscopically by the magnetization M, is given by Curie's law, Eq. 3-1, which may be written

$$C = \frac{M}{H} \cdot T,$$

when the ratio H/T is small. In the beginning, when $T = T_{max}$ (about $1°K$), a value of H is chosen at which Curie's law holds, and M is then measured with the aid of the coils shown in Fig. 3-8 and, of course, a suitable external circuit. Knowing M, H, and T, the Curie constant is calculated.

At the conclusion of the adiabatic demagnetization a very small

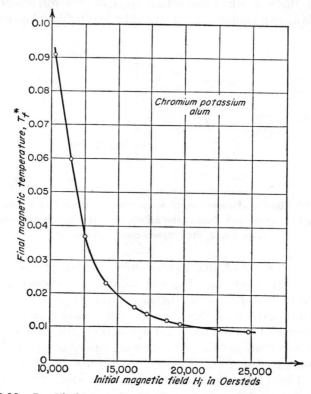

FIG. 3-11 **De Klerk's results in the adiabatic demagnetization of chromium potassium alum. (Initial temperature = 1.17°K.)**

field is applied (not enough to cause an appreciable rise of temperature), and the magnetization M is measured at this field. Knowing M, H, and C, a quantity called the *magnetic temperature* T^* is calculated by the *definition*

$$T^* = \frac{C}{M/H}. \tag{3-2}$$

Comparing the definition of T^* with Curie's law, it is seen that T^* is the Kelvin temperature if and only if the paramagnetic salt obeys Curie's law at the very low temperature achieved by the demagnetization process. Otherwise T^* may be entirely different from the true temperature. It is important therefore to understand how magnetic temperatures are converted to Kelvin temperatures.

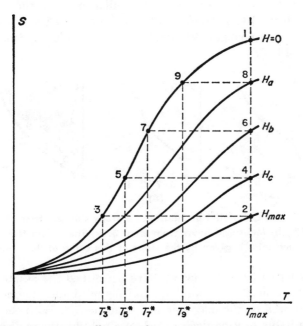

FIG. 3-12 Diagram to illustrate the measurements necessary to determine the values of T_3, T_5, T_7, and T_9 on the Kelvin scale.

§3-5 Conversion of magnetic temperature to Kelvin temperature. The first thing to do is to start at the same temperature but at different magnetic fields, demagnetize to zero field from each of these initial fields, and measure the corresponding magnetic temperatures. Experimental results obtained by de Klerk on chromium potassium alum are shown in Fig. 3-11. The initial temperature was 1.17°K in all cases. Note that the larger the initial field, the lower the final temperature.

The next step is to measure or calculate the entropy changes which occur during isothermal magnetizations from zero field to the various fields that were chosen as the starting fields for the demagnetizations. Consider the graph of Fig. 3-12. If the amounts of heat squeezed out during the processes $1 \to 2$ and $1 \to 4$ are measured, then subtracted, and then divided by the temperature T_{max}, the result is the entropy change from 2 to 4 *and also the entropy change from 3 to 5*. In similar manner the entropy changes from 3 to 7, from 3 to 9, and from 3 to 1 are obtained. These entropy changes, denoted by $S - S_3$, are plotted against T^* in Fig. 3-13 (a).

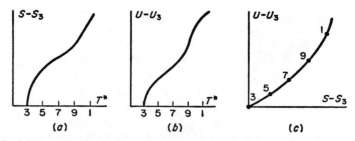

FIG. 3-13 **(a) Calculated or measured entropy changes from state 3 of Fig. 3-12. (b) Measured heats from state 3. (c) The Kelvin temperature at any point is given by the slope of this curve at that point.**

Now let us imagine that we have performed the adiabatic demagnetization 2 to 3 and are at the minimum temperature, characterized by the magnetic temperature T_3^*. With the aid of some suitable heating device (such as a gold heating coil used by Giauque, or absorption of gamma rays used by Kurti and Simon, or magnetic hysteresis used by de Klerk), let us measure the heat absorbed at zero magnetic field in going from 3 to 5, from 3 to 7, from 3 to 9, and from 3 to 1. Since these values of heat are measured in zero magnetic field, no work is done, and hence the heat is equal to the internal energy change $U - U_3$. These internal energy changes are plotted against T^* in Fig. 3-13(b).

Combining graphs (a) and (b), we get the graph of Fig. 3-13(c), in which $U - U_3$ is plotted against $S - S_3$. Since the point 3 is always the same, U_3 and S_3 are constants, and therefore the *slope* of this curve at any point is dU/dS, which according to Eq. 2-6 is the Kelvin temperature. Instead of turning back to Eq. 2-6, let us repeat the derivation. The first law says that

$$\text{energy change} = \text{heat} - \text{work,}$$

or

$$dU = dQ - dW.$$

But since $H = 0$, no work is done, or $dW = 0$ and also $dQ = T\,dS$. Hence

$$dU = T\,dS$$

and

$$T = \frac{dU}{dS}.$$

Once this procedure has been followed and a table of values of T^* and T (such as Table 3-3) has been obtained, other thermom-

TABLE 3-3   *Relation Between Magnetic and Kelvin Temperatures, Using Chromium Potassium Alum*

Magnetic temperature, T^*	Kelvin temperature, T
0.064	0.035
0.060	0.031
0.054	0.022
0.052	0.018
0.050	0.015
0.048	0.012
0.046	0.010
0.044	0.0088
0.042	0.0075
0.040	0.0065
0.038	0.0056
0.036	0.0047
0.034	0.0041
0.033	0.0039

eters such as radio resistors or germanium crystals can be calibrated. Nowadays, one can buy a paramagnetic salt mounted on a three-foot stainless steel tube to be inserted into a cryostat. A coil surrounding the salt is connected to a delicate bridge circuit with which the magnetization of the salt is measured. A calibration curve comes with the instrument.

The paramagnetic salt listed fourth in Table 3-2, cerium magnesium nitrate (CMN), is particularly noteworthy. Parallel to a certain axis the crystal is only slightly magnetic, with a Curie constant of only 0.0056, whereas perpendicular to this axis the crystal is about *sixty times as magnetic,* with a Curie constant equal to 0.318, and *obeys Curie's equation all the way down to* 0.006°K. Thus, starting at 1°K and a field of 7130 oersteds, the crystal needs merely to be rotated from the perpendicular to the parallel position for its temperature to drop below 0.01°K!

§3-6 **Superconductivity.** The complete loss of electrical resistance within a small temperature interval lying between 20°K and absolute zero is known as *superconductivity,* discovered in 1911 by Kamerlingh-Onnes. Of all the peculiar things that happen at low temperatures, superconductivity is:

(1) The most spectacular (persistent electric currents in metal rings should last over 100,000 years);

(2) The most useful to physicists and engineers (it is possible to make superconducting magnets, thermal switches, frictionless gyros, and small, fast, zero power-consuming computers); and

(3) The most challenging to theoretical physicists (superconductivity was unexplained for 46 years until Bardeen, Cooper, and Schrieffer in 1957 made the first theoretical breakthrough).

It is the purpose of this section to describe and explain enough of superconductivity to make clear how it is used in two devices employed in modern cryogenic equipment—superconducting magnets and thermal switches. A complete account of all the aspects of superconductivity would require a book of its own.

Superconductivity has been found to occur in about 23 elements, many alloys, and hundreds of compounds. In the case of an extremely pure, strain-free metal, the loss of resistance occurs within a temperature interval of the order of 0.01 degree, so that a definite temperature T_c can be attributed to the transition. The transition occurs more gradually in impure metals and in alloys and compounds, where it may require a temperature interval of as much as 2 degrees to cover the drop from normal resistance to zero resistance. When this is the case, the temperature at which the resistance is half its value in the normal state is often taken to be the transition temperature. Values of transition temperature T_c for a few useful superconductors are given in Table 3-4.

TABLE 3-4   *Properties of a Few Useful Superconductors*

Superconductor	T_c (°K)	H_0 (oersteds)
Hg	4.16	410
In	3.40	280
Nb	8.9	1960
Pb	7.22	812
Sn	3.74	307
Ta	4.38	860
Nb_3Sn	18.	200,000 to 300,000

No substance has yet been discovered with a transition temperature higher than that of the niobium tin compound, Nb_3Sn, which was discovered by Matthias at the Bell Telephone Laboratories. Transition temperatures are very temperamental. They depend on purity, strain, pressure, isotope content, and whether the material is in the form of a thin film. There are many superconducting compounds the component elements of which are not superconducting. Bismuth is not a superconductor under normal conditions, but becomes one at pressures between 20,000 and 40,000 atmospheres and also when in the form of a thin film.

When a normal material at room temperature is placed in a magnetic field, a certain density of magnetic lines of force measured by the *magnetic induction B* exists inside the material. If H is the intensity of the field, the ratio

$$\mu = \frac{B}{H}$$

is called the *permeability* of the substance. The paramagnetic salts listed in Table 3-2 have values of μ only slightly larger than 1. When B is less than H, a substance is said to be *diamagnetic,* and μ is less than 1. Bismuth is an example.

When a substance that becomes superconducting at low temperatures is placed in a magnetic field at room temperature, a magnetic field exists inside with a certain value of B. When the temperature is lowered below the transition point, and then the field is removed, what should happen to B? The answer is contained in Faraday's law of electromagnetic induction, namely,

$$\int E \cos \theta \, ds = -A \frac{dB}{dt},$$

where E is the electric intensity inside the superconductor, ds is an element of path, θ is the angle between E and ds, and A is the area. Since the resistance is zero, no electric field can exist inside because the electrons would be accelerated and, in moving faster and faster, the current would get larger and larger without limit. Since E must be zero, dB/dt is zero, and therefore

$$B = \text{constant (when } R = 0).$$

Physicists were so certain of this view that, in the twenty-year interval following the discovery of superconductivity, no one bothered to verify it. When Meissner performed the experiment in 1931 he discovered that B remained constant indeed, but at the value zero! In other words, when the material became superconducting it pushed all the magnetic lines *out* and preserved the value $B = 0$. Hence

$$B = 0 \text{ (when } R = 0).$$

Since $B = 0$, the permeability $\mu = 0$ and the material is not only diamagnetic, but *perfectly diamagnetic!* In determining whether a substance is superconducting and, if so, what is its transition temperature, one can measure either the resistance or the permeability.

Suppose a superconductor is maintained at a constant temperature *below* T_c and is placed in a magnetic field whose strength H can be varied at will. As H is increased, starting at zero, R or μ remains zero until a value of H is found at which R or μ will begin to rise. A further increase in H will restore the normal value of R or μ. The suddenness with which superconductivity is destroyed by the external magnetic field depends not only on impurities, strains, etc., but also on the shape and position of the sample in the magnetic field. In the case of a long, thin, single crystal placed longitudinally in a magnetic field, the transition takes place almost instantaneously at a value of H called the *threshold field*. When this experiment is repeated at a constant *lower* temperature, a larger magnetic field is found to be necessary to destroy superconductivity. Extrapolating to absolute zero, the threshold field at $T = 0$ is denoted by H_0; it is listed in Table 3-4.

Placing a superconductor in an external field of the proper magnitude and direction is not the only way to destroy its superconductivity. If the superconductor is fabricated in the form of a wire and connected to a battery, the resulting current in the wire provides its own magnetic field. When the current is big enough to make its magnetic field at the surface of the wire equal to the threshold field, superconductivity is destroyed. This is sometimes called the Silsbee effect. If a solenoid were wound with pure tin wire, a field of only a few hundred oersteds could be achieved, whereas with a solenoid wound with pure niobium wire one could get a few thousand oersteds.

The development in 1961 of methods of winding solenoids of Nb_3Sn (the substance is very brittle) has profoundly affected experimental procedures in low-temperature physics. A large electromagnet such as that depicted in Plate I with about 10 cubic inches of working space requires approximately 1.5 *million* watts of power. Most of this power goes into heat which has to be carried away with the aid of a cooling system of high capacity. The total cost of the magnet, the electrical generators, the cooling system, and control equipment is several hundred thousand dollars. By contrast, the cryostat housing the same working space for a superconducting magnet, providing an even larger magnetic field intensity, may occupy only a few cubic feet. It may weigh only a few hundred pounds and cost only a few tens of thousands of dollars. And as for power consumption, the needed power can be delivered by a storage battery with no heat developed in the magnet itself!

The range of magnetic fields and the temperatures at which these fields are available are shown in Fig. 3-14 for pure tin, pure niobium, and Nb_3Sn.

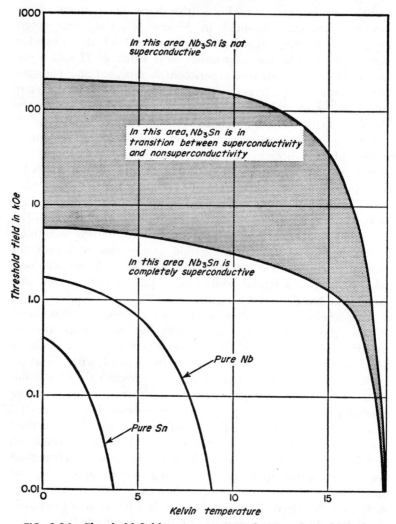

FIG. 3-14   Threshold field vs. temperature for Sn, Nb, and Nb₃Sn.

The value of H_0 for Pb given in Table 3-4 means that a field of 812 oersteds destroys the superconductivity of lead, no matter how low the temperature. If, therefore, a rod of lead is used to connect two bodies, and a coil of many turns of fine wire surrounds the rod, the superconductivity of the lead may be "turned off" by establishing a current of the proper magnitude, then turned on by opening the switch. Fig. 3-15 shows that, at all tem-

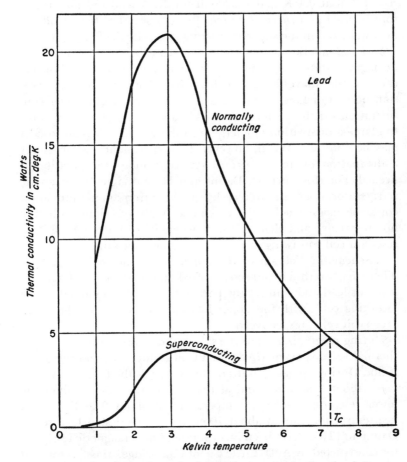

FIG. 3-15 The thermal conductivity of superconducting lead is very much less than that of normal lead.

peratures below 7.2°K, *superconducting lead is a poorer heat conductor than normal lead.* Below about 2°K, the heat conductivity of superconducting lead decreases as T^3 whereas that of normal lead decreases as the first power of T. At low enough temperatures, the heat conductivity of superconducting lead may become a very small fraction of that of normal lead; for example, it is $1/500$ at $0.3°K$ and $1/5000$ at $0.1°K$. A lead rod may therefore be used as a *thermal switch*. In the normal state it conducts heat well, and in the superconducting state, very badly.

§3-7 **Magnetic refrigerator.** At the conclusion of an adiabatic demagnetization, the temperature immediately starts to rise because of the unavoidable heat leak that is present in every cryostat, no matter how well designed. If the experiment under consideration can be performed quickly and does not itself involve too large a dissipation of energy into the system, adequate results may be obtained. When, however, the experiment itself requires a dissipation of about 50 ergs/sec, continuous refrigeration is needed. For this purpose, Daunt and Heer designed a magnetic refrigerator, operating in a cycle, in which iron ammonium alum could be magnetized while in contact with a helium bath (the hot reservoir) and demagnetized when in contact with a cold reservoir cell consisting of another paramagnetic salt.

A schematic diagram of the apparatus is shown in Fig. 3-16. There are two thin metal rods marked "thermal valve." These are made of lead and connect the paramagnetic salt contained in the "working cell" with the upper helium bath and with the lower "reservoir cell," respectively.

During magnetization of the salt in the working cell, the upper thermal valve is magnetized (good heat conductor), and the lower is not. During demagnetization, the reverse is the case. Each cycle takes about 2 minutes, and at the end of about 30 minutes the lower reservoir reaches a temperature of about $0.20°K$, with a heat extraction rate of about 70 ergs/sec. The design features of Daunt and Heer have been incorporated into a magnetic refrigerator constructed by A. D. Little Co. of Cambridge, Mass., composed of all necessary parts which perform automatically.

§3-8 **The polarization of magnetic nuclei.** The magnetism of the ionic magnets in a paramagnetic salt is due to uncompensated spins of the electrons which surround the nucleus. When the paramagnetic salt is placed in a magnetic field of the order of 10,000 oersteds and the temperature is lowered to about $1°K$ (so that the ratio H/T is of the order 10^4), the ionic magnets become partially oriented in the direction of the field. This situation is described by the word *polarization*. When the ionic magnets are

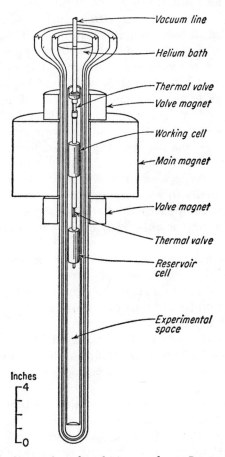

FIG. 3-16   **Magnetic cycle refrigerator due to Daunt and Heer.**

polarized, all the north poles point approximately in the *same* direction, although the orientation may not be perfect, that is, there may not be complete saturation.

When the axes of the ionic magnets are partially lined up

approximately parallel to the direction of the field but with about as many north poles as south pointing in the same direction, the magnets are said to be in *alignment*.

The particles inside the nucleus of an atom also have spins which give rise to *nuclear magnetism*. Nuclear magnets may also be polarized or aligned, but this requires a much larger ratio H/T, of the order of 10^7, because the *moment of a nuclear magnet is about one one-thousandth as large as that of an ionic magnet*. Even at a temperature of 0.01°K, a field of 100 kOe would be needed to provide a measurable polarization. Lining up nuclei all in the same direction, however, is a very valuable procedure for the nuclear physicist. If the nucleus is radioactive and emits alpha particles or beta particles or gamma rays, it is important to know whether these emanations are emitted as abundantly in one direction relative to the magnetic moment as in another. If they are emitted equally in all directions, one refers to *symmetry* or *isotropy* (accent on the second syllable). A preferred direction indicating asymmetry or anisotropy enables the nuclear physicist to:

(1) Test various *conservation laws* that are supposed to hold during nuclear disintegrations;

(2) Obtain information about the *shape* of the nucleus; and

(3) Obtain numerical values of certain constants needed in the theory of nuclear processes.

Nuclear polarization has been achieved by using a very low temperature and a very large field, but this procedure, known as the *brute force method,* has not been available in many laboratories.

The method of nuclear polarization that has been most fruitful was suggested first by Gorter and Rose and involves the behavior of a nucleus in the very strong *local* magnetic field (over 100 kOe) produced by *its* own electronic structure. To understand this, let us consider the implications of the theoretically derived and experimentally verified result that the *magnetization,* or measure of polarization of ionic or nuclear magnets, when they are so far from one another that they interact only weakly, *is a function of the ratio H/T*. If, say, 98 percent magnetic saturation is achieved with a particular paramagnetic salt at a temperature of 1°K with a field of 10,000 oersteds, then at a temperature of 0.01°K a field of only 100 oersteds would be needed to produce the same orientation of the ionic magnets. In other words, if the proper salt were used, starting at 1°K and demagnetizing from a field of 10,000 oersteds *not to zero field* but to a field of 100 oersteds, the temperature would drop to about 0.01°K, but the *orientation of the ionic magnets would still be about 98 percent.* One could also demagnetize to zero field, thereby reaching a somewhat lower temperature, and then raise the field to 100 oersteds and still have available about 98 percent polarization. Gorter and Rose realized that these polarized ionic magnets would give rise to a *unidirectional local field at the nucleus of each ion* that would be much larger than any field then achievable in the laboratory (before the advent of superconducting magnets), of the order of 100,000 oersteds or more.

This experiment was carried out several times by Roberts and his co-workers at Oak Ridge. They polarized Mn and Sm nuclei and detected their polarization by measuring their ability to scatter polarized slow neutrons from the Oak Ridge reactor.

A clever modification of the Gorter and Rose method was suggested by Bleaney, who pointed out that, when polarization is not needed, alignment can still be achieved without any external magnetic field by making use of the axial, nonhomogeneous electric field present in a single crystal. Experiments of this sort were performed by Bleaney and his co-workers at Oxford. The alignment of radioactive cobalt nuclei was detected by comparing the intensity of gamma-ray emission parallel to and at an angle to the alignment direction. There was a difference of as much as 40 percent at the lowest temperatures.

One of the most spectacular experiments in nuclear cryogenics was performed by Ambler, Hudson, and Wu at the National Bureau of Standards in Washington, in 1957. Lee and Yang, who were later awarded the Nobel prize, suggested that the nuclei of cobalt 60, in undergoing radioactive decay, might emit beta particles (electrons) more abundantly toward one magnetic pole than toward the other. To test this hypothesis, polarization of the cobalt nuclei was necessary. Beta-ray counters were then used to show whether there was a different reading in the direction of the north poles than in the reverse direction. The apparatus is shown in Fig. 3-17. The Co^{60}, obtained by neutron bombardment of nonradioactive Co^{59}, was introduced into the crystal lattice of cerium magnesium nitrate (CMN) in the form of a thin layer lying in the bottom of a cup-shaped housing of this material. The

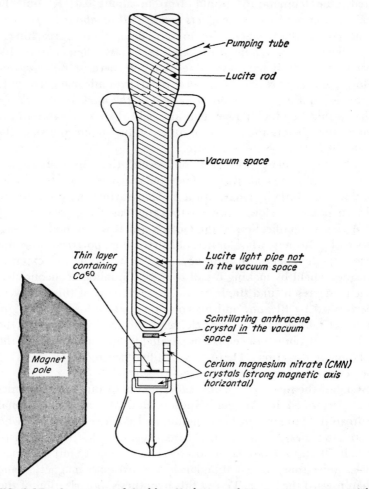

FIG. 3-17   Apparatus of Ambler, Hudson, and Wu to measure β-particle emission from polarized Co^{60} nuclei. (Liquid He and N_2 dewars and β-ray counters not shown.)

beta particles emitted by Co^{60} (in decaying to Ni^{60}) produced scintillations in a small crystal of anthracene. The light flashes traveled up a *light pipe* consisting of a four-foot lucite rod the upper end of which communicated with a photomultiplier tube and counter.

The strong magnetic axis of the CMN crystals was horizontal so that isothermal magnetization and adiabatic demagnetization to zero field could be accomplished with a horizontal magnetic field. By this means, the CMN housing and Co^{60} layer were cooled below 0.01°K. A solenoid was then slipped over the outer dewar, and a vertical field of about 100 oersteds was used to polarize the Co^{60} ions without appreciable warming of the CMN. The strong local fields at the nuclei of the ions polarized the Co^{60} nuclei, the direction of polarization depending upon the direction of the current in the solenoid. The north poles could therefore be made to face either toward the anthracene crystal or away from it.

The beta-particle ejection from Co^{60} is accompanied by gamma-ray emission. Although they are not shown in Fig. 3-17, there were two gamma-ray counters to detect and to measure any gamma-ray anisotropy. Since gamma-ray anisotropy had already been measured as a function of temperature, it served as a convenient thermometer. The curve in Fig. 3-18 shows that *there are many more beta particles emitted in a direction opposite the solenoid field,* that is, the south poles of the Co^{60} nuclei emit beta particles more abundantly than the north poles. This result is in direct contradiction to a nuclear principle, known as the *principle of the conservation of parity,* which had assumed that certain nuclear processes should behave the same way for one configuration of the nucleus as for its mirror image. The experimental proof that parity is not conserved in beta decay has had a profound effect on theoretical and experimental physics.

§3-9 **Production of still lower temperatures by nuclear demagnetization.** Since nuclear magnets are only about one one-thousandth as strong as ionic magnets, their polarization requires temperatures around 0.01°K and fields from 50 to 100 kOe. We have seen how local fields of this magnitude may be provided by the uncompensated spins of the electrons circulating outside of each nucleus itself. If these polarized nuclei could then be made to undergo a reversible adiabatic demagnetization, they would cool off to a temperature in the neighborhood of $10^{-5}°K$. Since, however, these nuclei occupy a thin layer on a large crystal at $0.01°K$, any loss of polarization would be much more isothermal than adiabatic. You cannot expect a few nuclei to cool off a big crystal.

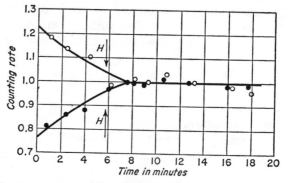

FIG. 3-18 Asymmetry in beta-particle emission from polarized Co⁶⁰ nuclei. Increasing time denotes increasing temperature. Maximum polarization is at the lowest temperature.

The only method that has been used so far to achieve temperatures below $10^{-3}°K$ involves a double process consisting of an ionic demagnetization followed by a nuclear demagnetization. Two separate magnetic fields supplied by two separate magnets are used, as shown symbolically in Fig. 3-19, a diagram prepared by Kurti of Oxford, in whose laboratory such experiments have been carried out. In each of the four parts of this figure the electronic stage represents a mass of chromium potassium alum in which are embedded 1500 enameled copper wires, each 3 ten-thousandths of an inch in diameter. The copper wires continue for a distance of about 8 inches and are then bent over and bound together to form the nuclear stage itself. The first part of the cooling is done with the aid of chromium ions and the second part by *copper nuclei*. The fine, insulated copper wires serve three purposes:

(1) As a heat-conducting medium between the nuclear and the electronic stages;

(2) To minimize eddy currents induced by demagnetization; and

(3) To produce a low temperature by nuclear demagnetization.
The four steps in Fig. 3-19 are as follows:

(a) Isothermal magnetization of the electronic stage;

(b) Adiabatic demagnetization of the electronic stage and cooling of the nuclear stage to 10^{-2}°K;

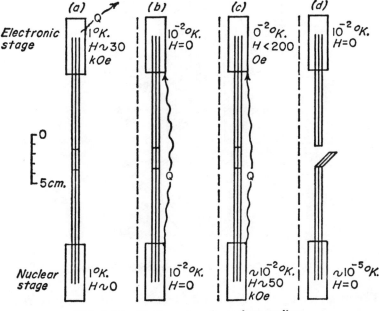

FIG. 3-19   The four steps in nuclear cooling.

(c) Isothermal magnetization of the nuclear stage; and

(d) Adiabatic demagnetization of the nuclear stage, with an accompanying temperature drop to about 10^{-5}°K.

The experiment is not as simple as it sounds. To quote Kurti, "The stringency of the conditions to be satisfied can be illustrated by remarking that even a minute amount of heating such as results from a small pin dropping through a height of one-eighth of an inch would warm a bulky specimen of several ounces from one-millionth of a degree to the starting temperature of one one-hundredth of a degree and thereby spoil the experiment." Even the eddy currents induced in the copper wires by virtue of slight variations of current (ripples) in the magnet coils must be prevented with the aid of the metal ripple shield shown in Fig. 3-20.

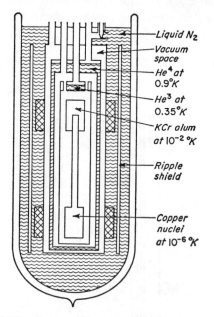

FIG. 3-20   Cryostat for nuclear cooling (symbolic).

The magnetic fields were supplied by solenoids in which currents of thousands of amperes were maintained. The complexity of the apparatus used at Oxford is shown in Plate II.

The lowest temperature ever obtained anywhere (1963) *is* 1.2×10^{-6}°K!

One of the biggest experimental difficulties to overcome in a double demagnetization process is the heat transfer between the nuclear and electronic stages. During the isothermal magnetization of the nuclear stage, this transfer must be good. During the following demagnetization, it must be poor. In the experiments of Kurti's group the fine copper wires represent a compromise which served both purposes only moderately well. Another difficulty is to separate the electronic from the nuclear stage by a big enough distance to confine each magnetic field to its own paramagnetic particles. Both of these problems can be partially solved by a clever method conceived by Blaisse. Suppose the nuclear stage constitutes a *core* completely surrounded by a crystal (or a group of crystals identically oriented) of cerium magnesium nitrate with its strong magnetic axis pointing toward, let us say, the *x*-axis. Suppose we perform the following operations:

(1) Magnetize isothermally at 1°K *in the x-direction.*

(2) Insulate thermally.

(3) *Rotate the field to the y-direction.* Since CMN is practically nonmagnetic in this direction, it therefore undergoes an adiabatic demagnetization and its temperature drops, *even though the field is still there.*

(4) Wait until the cold CMN has cooled the nuclear core, and then reduce the magnetic field to zero, thereby cooling the core by its own adiabatic demagnetization.

Another method, suggested by Kittel, is to polarize the nuclear magnets by microwaves at a temperature of 1°K and in a steady magnetic field of, say, 10^4 Oe. Removal of the high-frequency field adiabatically should result in cooling to about 10^{-3}°K. Then demagnetization from 10^4 Oe to zero should give about 10^{-7}°K. This method would have the advantage that there would be no heat transfer problems since all of the operations are made on the nuclear magnets.

§3-10 The third law of thermodynamics. We have seen how the Joule-Kelvin effect is employed to produce liquid helium at a temperature below 5°K. The rapid adiabatic vaporization of liquid helium then results in a further lowering of the temperature to about 1°K; the process of adiabatic demagnetization is used to lower the temperature of a set of ionic magnets to about 10^{-3}°K; and finally, a second adiabatic demagnetization of a set of nuclear magnets to about 10^{-6}°K.

The question that naturally arises is whether adiabatic demagnetization may be used to attain absolute zero. The clue to the answer to this question is provided by Fig. 3-21. In part (a) the two entropy-temperature curves, one for $H = 0$ and the other for any value of H, *do not meet at absolute zero.* That is,

$$S(0, 0) - S(0, H) > 0.$$

Under these circumstances, it is clear from Fig. 3-21(a) that a third adiabatic demagnetization from point 6 to point 7 would

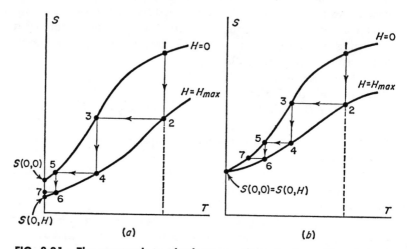

FIG. 3-21 The approach to absolute zero through successive adiabatic demagnetizations. (a) If $S(0,0)$ is different from $S(0,H)$, absolute zero may be reached. (b) If $S(0,0) = S(0,H)$, absolute zero is unattainable.

take the system to absolute zero. In part (b), on the other hand, the entropy-temperature curves have been drawn so that

$$S(0, 0) - S(0, H) = 0.$$

These curves show that no finite number of processes will result in the attainment of absolute zero. All experimental and theoretical evidence leads to the conclusion that the following two statements are equivalent, and are in agreement with the way nature behaves:

(1) *It is impossible by any procedure, no matter how idealized, to reduce any system to the absolute zero of temperature in a finite number of operations.*

This is known as either *the principle of the unattainability of absolute zero,* or, following Fowler and Guggenheim, *the unattainability statement of the third law of thermodynamics.*

TABLE 3-5 *The 100-Year Journey Toward Absolute Zero*

Date	Investigator	Country	Development	Temp., °K
1860	Kirk	Scotland	*First step toward deep refrigeration:* reached temperatures below freezing point of Hg.	234.0
1877	Cailletet	France	*First liquefied oxygen:* used throttling process from pressure vessel, obtaining fine mist only.	90.2
1884	Wroblewski & Olzewski	Poland	*First property measurements at low temperatures:* used small quantities of liquid N_2 and O_2.	77.3
1898	Dewar	England	*First liquefied hydrogen:* used Joule-Thomson effect and counterflow heat exchanger.	20.4
1908	Kamerlingh-Onnes	Netherlands	*First liquefied helium:* used same method as Dewar; shortly thereafter, lowered pressure over liquid to get 1°K.	4.2
1927	Simon	Germany & England	*Developed helium liquefier:* used adiabatic expansion from pressure vessel with liquid H_2 precooling.	4.2
1933	Giauque & MacDougall	U.S.	*First adiabatic demagnetization:* Principle first proposed by Giauque and Debye in 1926.	0.25
1934	Kapitza	England & U.S.S.R.	*Developed helium liquefier using expansion engine:* Made possible liquefaction of helium without liquid H_2 precooling.	4.2
1946	Collins	U.S.	*Developed commercial helium liquefier:* used expansion engines and counterflow heat exchangers.	2.0
1956	Simon & Kurti	England	*First nuclear cooling experiments:* used adiabatic demagnetization of nuclear stage of a paramagnetic salt.	10^{-5}
1960	Kurti	England	*Reached lowest temperature so far:* Nuclear cooling methods and apparatus recently refined.	10^{-6}

(2) *The entropy change associated with any isothermal, reversible process of a condensed system approaches zero as the temperature approaches zero.*

Let us call this theorem the Nernst-Simon statement of the third law of thermodynamics. Both this statement and the unattainability statement have had a long and chequered career since the original paper by Nernst in 1907. It took 30 years of experimental and theoretical research, during which time there were periods of great confusion, before all differences of opinion were resolved and the statement was agreed upon.

The fact that absolute zero cannot be attained is no cause for misgiving. A temperature of $3 \times 10^{-6}\,°K$ represents a fraction of room temperature, $300°K$, equal to

$$\frac{3 \times 10^{-6}}{3 \times 10^{2}} = 10^{-8}.$$

Cryogenics has therefore enabled us to get to one one-hundred-millionth of room temperature. The temperature of the sun, $6,000°K$, is only 200 times room temperature, and the temperature in the interior of the hottest star, about $3 \times 10^{9}\,°K$, is ten million times room temperature. Cryogenics is still ahead by a factor of ten.

A chronological account of the progress toward lower temperatures is given in Table 3-5, which is reproduced with the kind permission of *International Science and Technology*.

4. The Approach to Infinite Temperature

§4-1 **Filaments, furnaces, and flames.** The production of temperatures much higher than those occurring naturally represents a human activity that goes back many years. The ancient arts of heating homes, cooking food, illuminating rooms, ironing clothes, melting metals, and exploding old-fashioned bombs require the production of temperatures between 300°K and 3000°K. The synthesis of crystalline jewels, such as rubies, emeralds, and diamonds, requires somewhat higher temperatures. The processes and chemical reactions taking place under the conditions of a rocket engine exhaust or exploding TNT involve temperatures up to 6000°K.

The temperature range from 300 to 3000°K may be realized with the aid of the three devices shown schematically in Fig. 4-1:

(a) An electrically heated metal filament in a vacuum;

(b) A furnace in which a metal is heated by the large eddy currents induced in it by a high-frequency alternating current in an external coil surrounding the vacuum vessel; and

(c) A vacuum furnace whose source of energy is a beam of sunlight focused on a small area.

In all of these furnaces, the physical principle is the same: electromagnetic energy is transferred to a small body under conditions in which the loss of heat by conduction and convection is made as small as possible. At first, the rate of absorption of energy exceeds the rate of loss, and the temperature rises. After a while the rate of absorption is balanced by the rate of emission of radiant energy, and the temperature reaches a plateau which remains constant so long as the power input is constant.

Resistance thermometers have been used to measure temperatures up to 2000°K, and special thermocouples of refractory alloys

77

up to 3000°K, but these thermometers interfere with the production of high temperatures by conducting heat away at too great a rate. It is a much more accurate procedure to use the radiation issuing from the furnace as its own thermometer. How this is done will be explained in the next section.

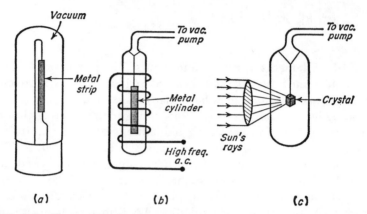

FIG. 4-1   Three methods of generating temperatures up to a few thousand degrees Kelvin: (a) electrically heated filament, (b) induction furnace, and (c) solar furnace.

Ordinary flames in which natural gas, an atomized liquid, or powdered coal react with the oxygen in the air rarely give temperatures higher than about 2000°K, but 4850°K may be achieved with the reaction

$$C_2N_2 + O_2 \rightarrow 2CO + N_2.$$

Other fuels such as H_2 and B_2H_6 combining with O_2, O_3, and F_2 give rise to flames whose temperatures go up to 6000°K. One of the hottest and most violent flames is produced by the reaction between hydrogen and fluorine,

$$H_2 + F_2 \rightarrow 2HF,$$

which takes place on contact, without the necessity for an ignition device or a catalyst. This reaction is being studied for possible use in rockets.

A 2000°K flame may be boosted to 5000° K by establishing a

high-voltage, low-current electric discharge across it. This is known as *electrical augmentation.*

Chemical reactions that take place at very high temperatures are complicated by the presence of many side reactions. According to Searcy, there are two fundamental laws of high temperature chemistry:

(1) At high temperatures, everything reacts with everything else; and

(2) The higher the temperature, the faster these reactions occur.

An explosion is a very fast reaction which involves a sharp rise in both temperature and pressure. When TNT explodes, the temperature may rise at first to a value around 5000°K.

§4-2 Planck's radiation law and the optical pyrometer. A substance may be stimulated to emit electromagnetic radiation in a number of ways:

(1) An electric conductor carrying a high-frequency alternating current emits radio waves;

(2) A hot solid or liquid emits thermal radiation;

(3) A gas carrying an electric discharge may emit visible or ultraviolet radiation;

(4) A metal plate bombarded by high-speed electrons emits X rays;

(5) A substance the atoms of which are radioactive may emit gamma rays; and

(6) A substance exposed to radiation from an external source may emit fluorescent radiation.

All these radiations are electromagnetic waves, differing only in wavelength. We shall be concerned here with thermal radiation only, i.e., with the radiation emitted by a solid or a liquid by virtue of its temperature. The radiation characteristics of gases require a special treatment. In the following pages we shall assume that gases are transparent to thermal radiation. When thermal radiation is dispersed by a suitable prism, a continuous spectrum is obtained. The distribution of energy among the various wavelengths is such that at temperatures below about 500°C most of the energy is associated with infrared waves, whereas at higher temperatures some visible radiation is emitted.

The higher the temperature of a body, the greater is the total rate of which energy is emitted.

The loss of energy due to the emission of thermal radiation may be compensated in a variety of ways. The emitting body may be a source of energy itself, such as the sun; or there may be a constant supply of electrical energy from the outside, as in the case of the filament of an electric light. Energy may be supplied also by heat conduction or by the performance of work on the emitting body. In the absence of these sources of supply, the only way in which a body may receive energy is by the absorption of radiation from surrounding bodies. In the case of a body that is surrounded by other bodies, the internal energy of the body will remain constant when the rate at which radiant energy is emitted is equal to that at which it is absorbed.

The rate at which a body emits thermal radiation depends upon the temperature and the nature of the surface. The total radiant energy emitted per unit time per unit area is called the *radiant emittance* of the body. For example, the radiant emittance of tungsten at 2450°K is 50 watt/cm². When thermal radiation is incident upon a body equally from all directions, the radiation is said to be *isotropic*. Some of the radiation may be absorbed, some reflected, and some transmitted. In general, the fraction of the incident isotropic radiation of all wavelengths that is absorbed depends upon the temperature and the nature of the surface of the absorbing body. This fraction is called the *absorptivity*. At 2450°K the absorptivity of tungsten is approximately 0.24. To summarize:

$$\text{Radiant emittance} = W = \frac{\text{total radiant energy emitted}}{\text{per unit time per unit area.}}$$

$$\text{Absorptivity} = a = \frac{\text{the fraction of the total energy of}}{\text{isotropic radiation that is absorbed.}}$$

There are some substances, such as lampblack, whose absorptivity is very nearly unity. For theoretical purposes it is useful to conceive of an ideal substance capable of absorbing all the thermal radiation falling upon it, so $a = 1$. Such a body is called a *blackbody*. A very good experimental approximation to a blackbody is provided by a cavity the interior walls of which are maintained at a uniform temperature and which communicates with the outside by means of a hole having a diameter small in comparison with the dimensions of the cavity. Any radiation entering the hole is partly absorbed and partly diffusely reflected a large number of times at the interior walls, so that only a negligible fraction eventually finds its way out of the hole. *This is true regardless of the materials of which the interior walls are composed.*

The radiation emitted by the interior walls is similarly absorbed and diffusely reflected a large number of times, so that the cavity is filled with isotropic radiation called *blackbody radiation*. Such radiation is studied by allowing a small amount to escape from a small hole leading to the cavity.

The fundamental tool of high-temperature thermometry is a blackbody at the melting temperature of gold, the so-called "gold point" of Table 1-2, 1336°K, as originally determined with the aid of a gas thermometer. The construction of the gold point blackbody used at the National Bureau of Standards is shown in Fig. 4-2. The energy needed to melt the gold is provided by an electric current in molybdenum windings.

The radiation from a blackbody at any temperature consists of a large amount of infrared, of wavelength larger than 7000 angstroms (1 angstrom = 10^{-10} meter). Since glass absorbs infrared radiation quite strongly, optical parts like lenses and prisms must be made of either quartz or rocksalt. Sometimes lenses are dispensed with entirely and all focusing is done with mirrors. If thermal radiation is focused on the slit of a spectrometer and is then dispersed into a spectrum with, let us say, a rocksalt prism, the spectrum may be examined by placing the thin junction of a thermocouple at various places, corresponding to various wavelengths. At any particular spot in the spectrum, the thermojunction intercepts a wavelength interval between λ and $\lambda + \Delta\lambda$. The thermocouple therefore measures the energy *within a wavelength band*. If this energy is divided by $\Delta\lambda$, the quotient is the energy per unit wavelength band. If we now measure the rate of emission and the area of the emitting surface, we will have finally *the*

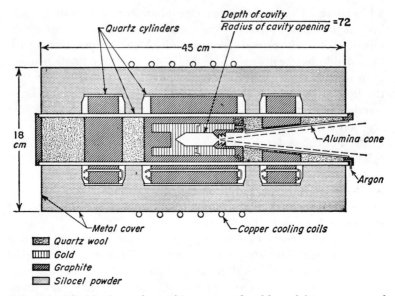

FIG. 4-2 **Blackbody at the melting point of gold, and furnace as used at the National Bureau of Standards.**

radiant emittance per unit wavelength band of a blackbody, $W_{B\lambda}$, where

$$W_{B\lambda} = \frac{\Delta W_B}{\Delta \lambda}.$$

Measurements of $W_{B\lambda}$ yield curves such as those shown in Fig. 4-3.

The first derivation of the relation between $W_{B\lambda}$ and λ at any value of T was given by Planck in 1900. Since this derivation involves the first use of the quantum hypothesis, it is of great historic interest. The following is an outline of the fundamental ideas.

Suppose that a string were stretched between two rigid supports one foot apart and plucked at some arbitrary point. The string would vibrate with many wavelengths at the same time: the fundamental vibration would have a wavelength of two feet, the second harmonic two halves of a foot, the third harmonic two-thirds of a foot, the fourth harmonic two-fourths of a foot, and the nth harmonic would have a wavelength of $2/n$. If you were asked to determine the number of standing waves whose wavelengths lie between 0.150 and 0.210 feet, you could find the answer merely by trying different values of n.

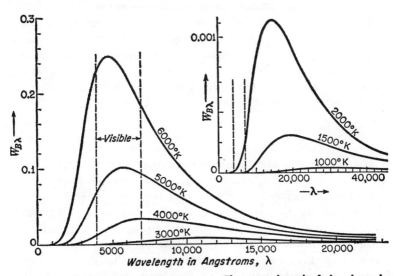

FIG. 4-3 Blackbody radiation curves. The wavelength λ is given in angstrom units ($1\text{Å} = 10^{-10}$m).

Suppose, however, that the string were capable of vibrating with hundreds or thousands of harmonics. These vibrations would have small wavelengths, and the difference in wavelengths between, say, the thousandth and the thousand and eighth would also be small. Suppose now you were asked how many vibrations there are with wavelengths lying between λ and $\lambda + \Delta\lambda$. The answer could be given if you could assume that the standing waves were so close in wavelength (almost a continuous spectrum) that the methods of calculus could be used to calculate $\Delta\lambda$. Thus, since the nth standing wave has a wavelength $\lambda = 2/n$, $\Delta\lambda$ (ignoring the minus sign caused by differentiation) is

$$\Delta\lambda = \frac{2}{n^2}\,\Delta n = \frac{\lambda^2}{2}\,\Delta n.$$

Hence,

$$\Delta n = \frac{2}{\lambda^2} \Delta\lambda.$$

Now consider the blackbody radiation in equilibruim with the interior walls of a constant temperature enclosure. This isotropic radiation may be regarded as an enormous number of standing waves, with wavelengths so close together that they constitute a continuous spectrum. If the condition for standing waves is applied, and proper account is taken of the three-dimensional character of the enclosure, the result is obtained that the number of vibrations whose wavelengths lie between λ and $\lambda + \Delta\lambda$ is

$$\frac{8\pi}{\lambda^4} \Delta\lambda. \qquad (4\text{-}1)$$

The next step is to compute the average energy associated with each vibration. Here is where the quantum theory first appeared. Instead of allowing a vibration to have any value of energy, Planck assumed that only whole number multiples of hc/λ are possible, where c is the speed of light and h is a universal constant, now known as *Planck's constant*. Assuming the standing waves of blackbody radiation to be equivalent to harmonic oscillators which interact with one another very weakly, and using the same statistical arguments that were used in Chap. 2, we may use the Boltzmann equation, Eq. 2-9, for the number of standing waves with energy u_i,

$$n_i = Ke^{-u_i/kT},$$

where $u_0 = 0$, $u_1 = hc/\lambda$, $u_2 = 2hc/\lambda$, and so on. The total energy U is given by $\Sigma n_i u_i$ and the total number N by Σn_i. The average energy $\bar{u}$ is therefore

$$\bar{u} = \frac{U}{N} = \frac{hc}{\lambda} \frac{\sum i e^{-ihc/kT\lambda}}{\sum e^{-ihc/kT\lambda}}.$$

When this expression is evaluated, it is found that

$$\bar{u} = \frac{hc/\lambda}{e^{hc/kT\lambda} - 1}. \qquad (4\text{-}2)$$

We now have the *number* of vibrations, Eq. 4-1, and the average energy per vibration, Eq. 4-2. Multiplying these two quantities gives us the radiant energy density whose wavelengths lie between λ and λ + Δλ, and from this we get

$$W_{B\lambda} = \frac{2\pi hc^2/\lambda^5}{e^{hc/kT\lambda} - 1}. \qquad (4\text{-}3)$$

This is the famous *blackbody radiation equation of Planck,* which, when plotted as in Fig. 4-3, gives curves that agree perfectly with experiment. Its truth is regarded as so certain that it is used to provide a set of rules for measuring temperatures above the gold point in a manner that is somewhat more precise than that shown in Fig. 4-4.

FIG. 4-4 The first optical pyrometer (From *Heat and Temperature Measurement,* by Robert L. Weber, © 1950 by Prentice-Hall, Inc., Englewood Cliffs, N.J.)

Suppose that an optical pyrometer such as that described in §1-2, and shown schematically in Fig. 1-5, is sighted on a blackbody at the gold point and the current necessary to make the lamp filament disappear is noted. Now suppose that another blackbody at a higher temperature T is sighted *with the same pyrometer at the same current setting*, but through a rapidly rotating disc equipped with a sector-shaped opening whose angle θ can be varied. The sectored disc transmits the same fraction $\theta/2\pi$ of the radiation of all wavelengths. The angle of opening can be varied until the radiation has the same intensity as that from the gold-point blackbody. When this is the case,

$$\frac{W_{B\lambda}(T_{Au})}{W_{B\lambda}(T)} = \frac{\theta}{2\pi},$$

and using the Planck equation,

$$\frac{e^{hc/kT\lambda} - 1}{e^{hc/kT_{Au}\lambda} - 1} = \frac{\theta}{2\pi}. \qquad (4\text{-}4)$$

The experssion hc/k, often represented by the symbol c_2, has the numerical value 1.438 cm°K. The wavelength transmitted by the filter is often chosen in the red region where $\lambda = 6.5 \times 10^{-5}$ cm (6500 angstrom units), and as mentioned before, $T_{Au} = 1336$°K. Solving Eq. 4-4 for T gives the temperature of the unknown blackbody.

In this way, a second optical pyrometer may be calibrated so that a value of the lamp current at which the filament disappears tells the temperature of a blackbody at which the pyrometer is sighted. If the red filter transmitted only a very narrow band of wavelengths and if all unknown bodies whose temperatures must be measured were blackbodies, everything would be fine, and no further physical ideas or instruments would be needed. This, however, is not the case. Elaborate and difficult corrections have to be made to take into account the finite band width of the filter and the departure from blackbody conditions.

§4-3 **Arcs and plasmas.** The very intense light needed to illuminate a moving-picture film in order to cast a bright image on a large screen far away from the projector is usually provided by the crater in the positive electrode of a dc carbon arc. In large theaters as much as 100 amperes is sent through the arc, and a temperature of 4000°K may be achieved. At higher currents, the temperature may be high enough to evaporate the electrodes. Replacing the positive carbon by a tungsten rod and using a copper plate as the negative electrode, and maintaining a current of 500 amperes for a short time, a temperature of 30,000°K has been achieved. The arc itself consists of electrons, positive ions of nitrogen and oxygen from the air, and ions of the material of the electrodes. The more this mixture of gases spreads, the more readily will the electrodes and the arc housing melt or evaporate and the more limited will be the temperature that can be achieved.

The Weiss arc, built at the University of Kiel in Germany in 1954, achieves a much higher temperature by using a vortex of water to cool the electrodes and container, to absorb radiation, and to constrict the arc stream to a region along the axis of the vortex. The water pressure provides the preliminary constriction, and the electromagnetic "pinch effect" does the rest. The pinch effect is merely the attraction of parallel electric currents. Two parallel wires near each other carrying currents in the same direction attract each other with a force that varies inversely with the distance. Once parallel streams of ions and electrons get close enough together, they contract further. The power input in the Weiss arc is therefore constricted to a small space, and with 1200 amperes a temperature of 50,000°K has been achieved.

At temperatures in the 20,000 to 50,000°K range, there are no solids or liquids. Everything has been vaporized; all molecules have been dissociated into atoms, and many of the atoms have been either ionized or raised to higher energy states. The mixture of atoms, ions, and electrons is called a *plasma,* and if the plasma is caused to issue from the arc in the form of a rapid stream, the stream is called a *plasma jet.* If the jet is used for melting solids, boring holes, cutting slabs, etc., it is called a *plasma torch.* Almost all plasma generators are based on the principle of the Weiss arc. Modern generators are much larger, and many use a constricting vortex of gas instead of water.

A schematic diagram of a plasma generator is shown in Fig. 4-5, and a picture of a plasma jet in Plate III.

The methods of optical pyrometry may be valuable in measuring the surface temperature of a plasma jet, but give no clue to the interior temperature of the generator since a pyrometer is not calibrated in this range. It is therefore worthwhile to explain how a knowledge of the degree of ionization (the fraction of atoms that are ionized) may be used to measure the temperature. The theory is due to Saha and is based on well-known equations of thermodynamics. Consider a homogeneous gas mixture consisting

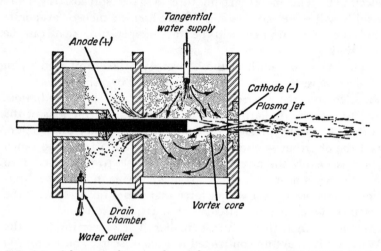

FIG. 4-5　Schematic diagram for a plasma generator (14,000°K). Water-stabilized DC arc.

of atoms (symbol A), positive ions formed by removing one electron (symbol A^+), and electrons (symbol e), in a vessel at temperature T and pressure P, in thermodynamic equilibrium. The reaction is written

$$1A \rightleftharpoons 1A^+ + 1e$$

and, at equilibrium, the *law of mass action* says that

$$\ln \frac{x_{A^+} \cdot x_e}{x_A} P = \ln K, \qquad (4\text{-}5)$$

where the x's stand for mole-fractions and K is a function of temperature only. If we imagine that there were n_0 moles of atoms at the start and that at equilibrium a fraction ϵ of these become ionized, then

PLATE I This huge magnet is the famous electromagnet of the Laboratoire Aimé Cotton at Bellevue, near Paris. It was used by Simon and Kurti before World War II. (Courtesy of N. Kurti, F.R.S., Clarendon Laboratory, Oxford University.)

PLATE II Apparatus used at Oxford for the production of temperatures in the microdegree range. (Courtesy of N. Kurti, F.R.S., Clarendon Laboratory, Oxford University.)

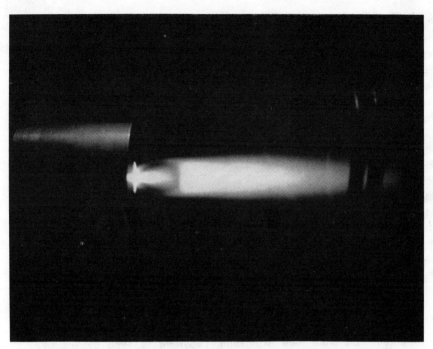

PLATE III Plasma jet in nitrogen. (Courtesy of Warner and Swasey Control Instrument Division, Flushing, N.Y.)

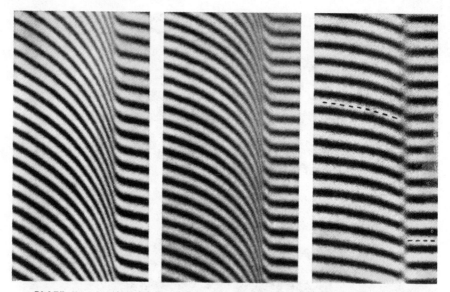

PLATE IV   Mach-Zehnder fringes, photographed by Walker Bleakney, showing three stages in the formation of a shock wave. The displacement of an interference fringe from the horizontal position is a measure of the density. (Reprinted with permission. Copyright © 1963 by Scientific American, Inc. All rights reserved.)

n_A = number of moles of atoms at equilibrium = $n_0(1 - \epsilon)$,

n_{A^+} = number of moles of ions at equilibrium = $n_0\epsilon$,

n_e = number of moles of electrons at equilibrium = $n_0\epsilon$,

Adding these three n's, we get for the total number of moles the quantity $n_0 (1 + \epsilon)$. The mole-fractions are obtained by dividing each n by the total number of moles. Hence,

$$x_A = \frac{1 - \epsilon}{1 + \epsilon}, \quad x_{A^+} = \frac{\epsilon}{1 + \epsilon}, \quad x_e = \frac{\epsilon}{1 + \epsilon},$$

and Eq. 4-4 becomes

$$\ln \frac{\epsilon^2}{1 - \epsilon^2} P = \ln K.$$

Expressing $\ln K$ in terms of temperature and atomic constants, Saha obtained the equation

$$\log \frac{\epsilon^2}{1 - \epsilon^2} P = -5050 \frac{E}{T} + \frac{5}{2} \log T + \log \frac{\omega_{A^+}\omega_e}{\omega_A} - 6.491$$

(4-6)

where log means the common logarithm, P is measured in atmospheres, E is the ionization potential in volts, and ω is a small integer depending on the nature of the lowest energy level and called its statistical weight. As an example of the use of Saha's equation, consider the case of cesium, which has a particularly low value of E (3.87 volts), and where $\omega_A = 2$, $\omega_{A^+} = 1$, and $P = 10^{-6}$ atm. A graph of Eq. 4-6 for cesium is shown in Fig. 4-6. A measurement of the degree of ionization of cesium is therefore a convenient way of measuring temperature, at least in the range from 1900 to 3000°K. With other atoms and at other pressures, higher temperature ranges are available. At higher temperatures, the Saha equation must be generalized to allow for multiple ionization, that is, ions formed by removing more than one electron from a neutral atom.

If the tube containing the mixture of atoms, ions, and electrons is equipped with two electrodes across which a potential difference is applied, the current registered by an external galvanometer depends upon the degree of ionization. If suitable calibration data are at hand, this is the simplest method of measuring ϵ, and then of finding the temperature. The curve shown in Fig. 4-7 gives a calculated value of the electrical conductivity of an argon plasma as a function of temperature.

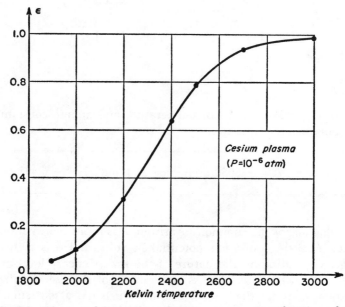

FIG. 4-6 **Degree of ionization of cesium vapor as a function of temperature.**

§4-4 Spectral lines. The narrow slit marked S in Fig. 4-8(a) when illuminated by the red light from a cadmium lamp gives rise to an image on the screen W when the lens L is uncovered. Suppose a diaphragm D is used to limit the aperture of the lens. How does the image of the narrow slit depend on the aperture a? This may be easily demonstrated experimentally. If the aperture a is very narrow like the source slit S, the image of S on the screen is not determined by geometrical optics, but consists of a diffraction pattern which is broader the smaller the value of a, as shown in Fig. 4-8(b). If, however, we make *a* as wide as the lens itself, say, over an inch, then the image of the narrow slit *S* is extremely sharp, and diffraction will cause no trouble.

Suppose now that the entire light from a cadmium lamp is used

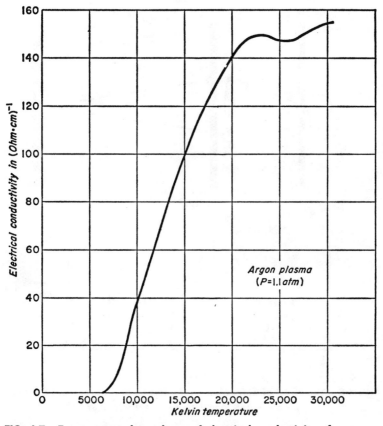

FIG. 4-7   **Temperature dependence of electrical conductivity of an argon plasma.**

to illuminate the very narrow slit of the prism spectrograph shown in Fig. 4-9 where all lenses are large. A number of sharp images of the spectrograph slit appear in the focal plane of the telescope lens, one for each color in cadmium light. Between the colored slit images will be darkness. The cadmium vapor lamp is said to emit a *line spectrum,* in contrast with the continuous spectrum emitted by an incandescent solid or liquid.

No matter how narrow the spectrograph slit is made, any one colored image on the screen has a definite width, depending upon

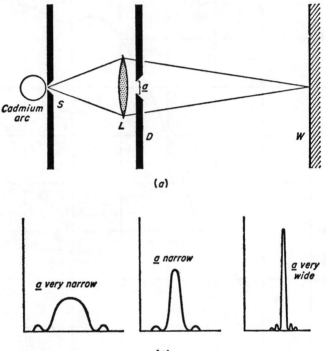

FIG. 4-8   The effect of diffraction on the image of a narrow slit S formed by a covered lens (L and D) on a screen W.

the wavelength spread of the spectral line itself. A spectrum line, in other words, is not infinitesimally narrow at one and only one wavelength, but consists of a distribution of energy among a band of wavelengths, as shown in Fig. 4-10, where a spectral line of simple shape is depicted. The spread of wavelengths may be due to many factors, such as:

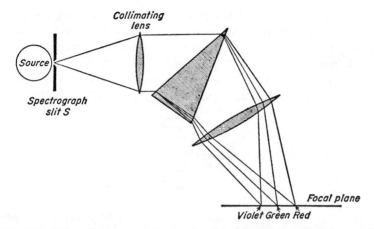

FIG. 4-9 When there are distinct colored images of the slit S on a dark background, the source is said to emit a line spectrum.

(1) The changes in wavelength arising from the motions of the emitting atoms, known as the *Doppler effect;* or

(2) The changes in wavelength due to interruptions in the process of emission by virtue of *atomic collisions.*

The distribution of wavelengths in a given spectral line is specified by giving the wavelength spread at half-maximum, shown in Fig. 4-10 and labeled *half-breadth.*

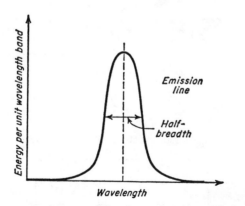

FIG. 4-10 Even a single spectral line has a finite spread of wavelengths.

When the pressure of the gas composed of emitting atoms is low, the collision broadening is negligible, and the half-breadth is due almost entirely to Doppler broadening. If an atom, capable of emitting a wave of wavelength λ_0 while stationary, is moving away from an observer with a velocity whose component along the line of sight is u, the wavelength seen by the observer is λ, where

$$\frac{\lambda - \lambda_0}{\lambda_0} = \frac{u}{c},$$

c being the speed of light. Hence

$$u = c\frac{\lambda - \lambda_0}{\lambda_0},$$

The probability that the speed of an atom of mass m lies between u and $u + du$ is proportional to

$$e^{-\frac{mu^2}{2kT}} du = e^{-\frac{Mu^2}{2RT}} du,$$

where M is the molecular weight and R is the universal gas constant. The intensity distribution, therefore, of a Doppler-broadened line is obtained by substituting the Doppler expression for u. Hence the distribution is given by

$$I = I_0 e^{-\frac{Mc^2}{2RT}\left(\frac{\lambda-\lambda_0}{\lambda_0}\right)^2}.$$

At half maximum $I/I_0 = \frac{1}{2} = e^{-0.69}$. Hence

$$\frac{Mc^2}{2RT}\left(\frac{\lambda - \lambda_0}{\lambda_0}\right)^2 = 0.69,$$

$$2\frac{\lambda - \lambda_0}{\lambda_0} = 2\sqrt{\frac{0.69 \times 2RT}{Mc^2}},$$

$$= \frac{2.14 \times 10^4}{c}\sqrt{\frac{T}{M}},$$

and finally, the Doppler breadth, $\Delta\lambda_D$, is given by

$$\boxed{\frac{\Delta\lambda_D}{\lambda_0} = 7.16 \times 10^{-7}\sqrt{\frac{T}{M}}.} \qquad (4\text{-}7)$$

Therefore a measurement of the wavelength at the center of the line λ_0, the molecular weight of the gas M, and the Doppler breadth $\Delta\lambda_D$ enable one to calculate the temperature T.

An example of this procedure is provided by the measurements of Griffin, McNally, and Werner on a spectral line emitted by doubly ionized carbon atoms in an intense arc produced at the Oak Ridge National Laboratory. A current of 158 amperes was

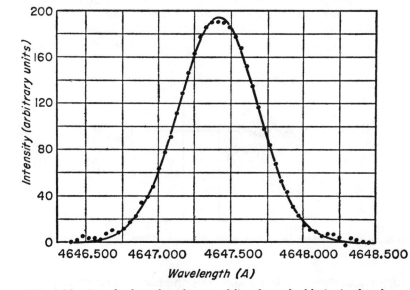

FIG. 4-11 Doppler-broadened spectral line from doubly ionized carbon. (P. M. Griffin, J. R. McNally, Jr., and G. K. Werner, *Temperature*, Vol. 3, Part I, p. 659, Reinhold Publishing Corp., 1962.)

maintained in an magnetically constricted carbon arc. The potential difference was 97 volts so that the power input was over 15,000 watts. The graph of intensity versus wavelength is shown in Fig. 4-11, from which we get $\Delta\lambda_D = 0.651 \times 10^{-8}$ cm and $\lambda_0 = 4647 \times 10^{-8}$cm, so that

$$\frac{\Delta\lambda_D}{\lambda_0} = 1.4 \times 10^{-4}.$$

Since

$$T = \frac{\left(\frac{\Delta\lambda_D}{\lambda_0}\right)^2 M}{51.3 \times 10^{-14}},$$

and M is 12, we get the result that

$$T = 458,000°K.$$

According to simple atomic theory, a spectral line is emitted when an atom, in an upper or *excited* energy level, undergoes a transition to a lower level. Of the three atomic energy levels shown in Fig. 4-12 the lowest, marked 0, is the normal state of the

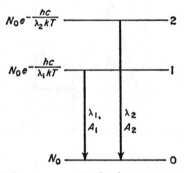

FIG. 4-12 Spectral lines are emitted when atoms undergo transitions from upper to lower energy levels.

atom, and those marked 1 and 2 are the nearest excited states. If the transition between 1 and 0 gives rise to the spectral line of wavelength λ_1, then the energy difference between 1 and 0, $u_1 - u_0$, is given by

$$u_1 - u_0 = \frac{hc}{\lambda_1}.$$

Similarly,

$$u_2 - u_0 = \frac{hc}{\lambda_2}.$$

If there are N_0 atoms in the normal state, then the number of atoms in each of the excited states is given by the Boltzmann equation,

$$\frac{N_1}{N_0} = e^{-\frac{u_1 - u_0}{kT}} = e^{-\frac{hc}{\lambda_1 kT}},$$

$$\frac{N_2}{N_0} = e^{-\frac{u_2 - u_0}{kT}} = e^{-\frac{hc}{\lambda_2 kT}},$$

as indicated in Fig. 4-12.

The energy emitted per unit time, or the *intensity* of a spectral line, depends upon four factors:

(1) The number of atoms in the upper energy level;

(2) The energy of the photon emitted;

(3) The probability that the atom will undergo the transition from upper to lower level per unit time; and

(4) The statistical weights of the two states, ω_1 and ω_2.

If A_1 is the probability per unit time of the transition from 1 to 0, A_2 that from 2 to 0, then

$$I_1 = \frac{\omega_1 A_1 hc}{\lambda_1} N_0 e^{-\frac{hc}{\lambda_1 kT}},$$

$$I_2 = \frac{\omega_2 A_2 hc}{\lambda_2} N_0 e^{-\frac{hc}{\lambda_2 kT}}.$$

The ratio of the intensities is

$$\boxed{\frac{I_1}{I_2} = \frac{\omega_1 A_1 / \omega_2 A_2}{\lambda_1 / \lambda_2} e^{-\frac{hc}{kT}\left(\frac{1}{\lambda_1} - \frac{1}{\lambda_2}\right)}.} \qquad (4\text{-}8)$$

There are many ways of measuring the probabilities A_1 and A_2, which are known as the *Einstein A-coefficients*. It can be shown that the sum of all the A's corresponding to all the ways of going from *one* upper state to all the lower states is the reciprocal of the average time that the atom spends in the upper state, or *lifetime of the excited state.* Many measurements of life-times and also many theoretical calculations have been made so that we are in possession of quite a few values of the Einstein A coefficients. It follows therefore that, if two spectral lines whose A coefficients are known are compared in intensity, everything in Eq. 4-8 is known except T, which therefore can be calculated.

An example of this procedure is provided by the work of W. E. Hill, who studied the relative intensity of the two copper lines 5153 and 5700. Eq. 4-8 becomes in this case

$$\frac{I(\text{Cu }5153)}{I(\text{Cu }5700)} = 590e^{-\frac{26,200°\text{K}}{T}},$$

which is plotted in Fig. 4-13.

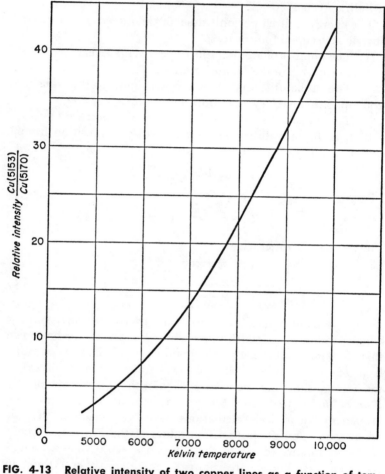

FIG. 4-13   **Relative intensity of two copper lines as a function of temperature (W. E. Hill).**

To summarize, these are the three kinds of measurements that are usually made on spectral lines:

(1) A spectral line emitted by an ionized atom is compared in intensity with that from an un-ionized atom and the degree of ionization ϵ of a gas is obtained. Knowing the pressure of the gas P and the ionization potential E, we may obtain the temperature with the aid of Saha's equation

$$\log \frac{\epsilon^2}{1 - \epsilon^2} P = -5050 \frac{E}{T} + \frac{5}{2} \log T + \text{constant.}$$

We may call this temperature the *"ionization temperature."*

(2) The Doppler width $\Delta\lambda_D$ of a spectral line is measured when the line is emitted by atoms of a gas whose atomic weight is M. The temperature is calculated from the equation:

$$\frac{\Delta\lambda_D}{\lambda_0} = 7.16 \times 10^{-7} \sqrt{\frac{T}{M}}.$$

This is the *"Doppler temperature."*

(3) The ratio of the intensities of two lines of wavelengths λ_1 and λ_2 from the same atom with Einstein A coefficients A_1 and A_2 and statistical weights ω_1 and ω_2 is measured and the temperature obtained from

$$\frac{I_1}{I_2} = \frac{\omega_1 A_1/\omega_2 A_2}{\lambda_1/\lambda_2} e^{-\frac{hc}{kT}\left(\frac{1}{\lambda_1} - \frac{1}{\lambda_2}\right)}.$$

This is the *"excitation temperature."*

Suppose that all three methods are employed with the same gas. The three values of T are identical if the gas is in thermodynamic equilibrium. If, however, the gas is moving rapidly, and its atoms undergo frequent collisions with electrons whose velocitiy is maintained by a heavy electric discharge in the gas, and the temperature is nonuniform, then it is quite possible for the three values of temperature to differ. Then we have to be content with three "kinds of temperature." Such is the case with a rapidly moving plasma. When a plasma is far from equilibrium, the very *concept of temperature itself has little meaning,* for temperature is a property that determines thermal equilibrium between systems or parts of systems. When equilibrium is absent, the quantities that we have been calling "temperatures" are only parameters calculated from specific properties of the system—the degree of ionization, the kinetic energy distribution of atoms, or the populations of energy levels.

§4-5   **Shock waves.** Sound is a physiological sensation attending the reception by the ear of longitudinal waves in the frequency range from about 20 cycles per second to about 20,000 cycles per second. These waves are familiarly called *sound waves*. They are characterized by the fact that (1) they have very small amplitudes and (2) they have a speed that is independent of the wavelength, but dependent on the temperature. When the medium is a gas at low enough pressure to obey the ideal gas law, the speed v is given by

$$v = \sqrt{\frac{\gamma RT}{M}},$$

where γ is the ratio of the specific heat at constant pressure to the specific heat at constant volume, R is the universal gas constant, and M is the molecular weight.

Regarding air at atmospheric pressure as an ideal gas, we have at room temperature:

$$T = 300°K,$$
$$M = 28.8 \text{ gm/mole},$$
$$R = 8.31 \times 10^7 \text{ ergs/mole.°K},$$
$$\gamma = 1.40 \text{ (because } N_2 \text{ and } O_2 \text{ are diatomic)}.$$

Hence

$$v = \sqrt{\frac{1.40 \times 8.31 \times 10^7 \frac{\text{dyne} \cdot \text{cm}}{\text{mole} \cdot °K} \times 300°K}{28.8 \frac{\text{gm}}{\text{mole}}}} = 3.48 \times 10^4 \frac{\text{cm}}{\text{sec}}.$$

Longitudinal waves that produce even the loudest sounds involve very small *pressure amplitudes,* that is, departures from atmospheric pressure. For example, while atmospheric pressure is very nearly 10^6 dynes/cm², the pressure amplitude of the loudest tolerable sound is only 280 dynes/cm².

A shock wave is a moving pressure disturbance in which the pressure amplitude is very much larger than that in a sound wave, and whose velocity is much greater than that of a sound wave.

The ratio of a velocity to that of a sound wave is called the *Mach number*. A speed twice that of sound is said to be "Mach 2." Shock waves have been produced up to Mach 200. The "sonic boom" that precedes the sound of supersonic aircraft carries a large amount of energy. Under controlled conditions, this quantity of energy may be used to produce momentarily in a small volume of gas a temperature change that is greater than that produced by any other agency, with the exception of nuclear explosions. To understand how this is done, let us first describe a simple shock tube and the nature of the measurement made at various places in the shock tube.

A shock tube consists of a long pipe a few square inches in cross section, equipped with a breakable diaphragm which separates the pipe into two regions, in one of which there is a light high-pressure gas and in the other a low-pressure test gas. When the diaphragm is ruptured, a great discontinuity of pressure is produced at the interface between the light gas and the test gas. As a result, the test gas near this interface is compressed adiabatically and undergoes a large temperature rise. In the meantime, the high-pressure gas, acting like a piston, is driving ahead, but the high temperature pulses behind it are catching up because $v \propto \sqrt{T}$. They all catch up to form a large pressure pulse which moves through the gas with speed greater than sound. These processes are indicated roughly in the four graphs shown in Fig. 4-14. The final pressure "wall" represents an extremely sharp, momentary pressure change which is almost a discontinuity. The measurement of such a rapid process taking place in such a small space requires the use of an optical interferometer.

If a beam of light issuing from one point of a source is split into two parts, and the parts are then brought together after each part has traveled a different distance, the two beams may arrive in phase, producing "constructive interference" (brightness), or in opposite phase, producing "destructive interference" (darkness), or in some condition between these two extremes. On a screen or a photographic plate where many such beams of light come together there are regions of brightness shading off into darkness and then repeating the variations again and again. These gradations are called *interference fringes*, and the device for splitting beams of light into parts and bringing them together again is called an *optical interferometer*. There are over a dozen different types of optical interferometers, each designed for a different purpose. For example, a Michelson interferometer is often used to measure a distance in wavelengths of the orange-red light of Krypton 86, the light now used to define the length of the meter;

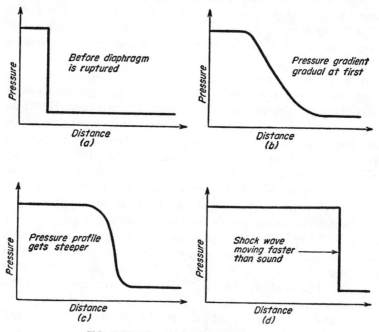

FIG. 4-14   Formation of a shock wave.

a Lloyd's mirror interferometer may be used to show that a beam of light undergoes a phase shift upon reflection from glass.

The interferometer that has proved most useful in shock tube measurements is known as the *Mach-Zehnder* interferometer; it is depicted in Fig. 4-15. The two interfering beams of light are formed at the first beam splitter. One goes through the compensators, which are two thicknesses of glass identical with those of the two windows of the shock tube. The other goes through the shock tube. If there were no gas in the shock tube, the interference

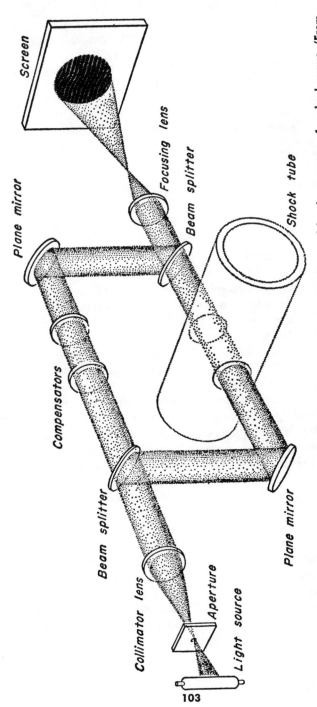

FIG. 4-15 The Mach-Zehnder interferometer indicates changes in gas density caused by the passage of a shock wave. (From Malcolm McChesney, "Shock Waves and High Temperatures," *Scientific American*, Feb. 1963. Reprinted with permission. Copyright © 1963 by Scientific American, Inc. All rights reserved.)

103

fringes would be perfectly straight. When gas is allowed to flow in slowly, the fringes shift, and the number of fringes that move across the field of view is proportional to the gas density. Thus one can calculate the density once the fringe shift is measured. The ideal gas law may be written in the form

$$PV = \frac{m}{M} RT,$$

where m is the mass of gas and M is the molecular weight. Since the density $\rho = m/V$,

$$T = \frac{M}{R} \frac{P}{\rho},$$

from which the temperature may be calculated when the pressure P and the density ρ are known.

The beautiful photographs shown in Plate IV, made by Walker Bleakney at Princeton University, show the formation of a shock front. The fringe shift in the third picture is of 7 fringes, as shown by the dotted lines.

Shock waves have been produced in argon with accompanying temperatures of 20,000°K or more. Under these conditions argon shows more than 50 percent ionization. All gases in this temperature range become excellent electric conductors. For example, with an electric field of one volt per centimeter, a current density of 100 amperes per square centimeter has been obtained. Nowadays, temperatures in modern laboratory shock waves are aproaching the thermonuclear range—around a million degrees Kelvin.

One of the most effective ways of producing a shock wave of enormous speed and high temperature makes use of the high conductivity of a plasma. A plasma arc is illustrated in Fig. 4-16. By discharging a heavily-charged bank of capacitors through the arc, a momentary current of about one million amperes is produced for about one millionth of a second. The circular magnetic field surrounding the arc discharge holds it together by the pinch effect, whereas the contrary circular magnetic field outside the arc produces a strong repulsion which starts the shock wave down the tube. At Mach 200, the temperature in the shock wave is about 500,000°K.

§4-6 **Fission of uranium.** The Planck radiation law and the Saha ionization equation as used by astronomers provide fairly reliable values of the surface temperature of stars, that is, the temperature of the outer layers of gas, such as the chromosphere of our sun. These temperatures are only in the thousands or tens of thousands of degrees. Astronomers have been able to make

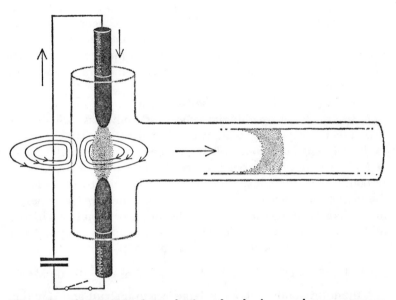

FIG. 4-16 Electromagnetic production of a shock wave by an enormous momentary current caused by the discharge of capacitors through a plasma. (From Malcolm McChesney, "Shock Waves and High Temperatures," *Scientific American*, Feb. 1963. Reprinted with permission. Copyright © 1963 by Scientific American, Inc. All rights reserved.)

estimates of the temperature in the interior of stars by using the ideal gas equation and by taking into account the opposing effects of gravitational contraction and radiation pressure. On the basis of such calculations, it has been known for many years that the internal temperatures of stars must be in the ten million degree range or higher. Of course, this temperature cannot be measured, since no radiation comes *directly* from the center of a star. The calculated value, however, fits in with every astronomic fact and with the conclusions of modern nuclear physics.

Ever since the explosion of the first uranium bomb in New Mexico in 1945, physicists have been able to develop temperatures as high as those inside the stars and have even duplicated, although not in a steady controlled way, some of the nuclear reactions that take place inside the stars. Someday we shall learn to control these processes and to use them to generate electric power. The study of the production and measurement of temperatures of the order of ten million degrees to a hundred million degrees is therefore now an important part of modern physics and engineering.

When a uranium nucleus absorbs a neutron and undergoes fission, two nuclei of roughly the same mass, known as *fission products*, are formed along with two other neutrons, some gamma radiation, and an amount of kinetic energy equal to 200 MeV. (1 MeV per uranium atom is about 10^8 calories per gram.) Corresponding to different specific modes of break-up, there are many different fission products—in fact, over one hundred isotopes of more than 20 different elements. All of these atoms are in the middle of the periodic table, with atomic numbers ranging from 34 to 58. A typical fission reaction is as follows:

$$_{92}U^{235} + _0n^1 \rightarrow _{92}U^{236} \rightarrow _{54}Xe^{140} + _{38}Sr^{94} + 2_0n^1 + \gamma + 200 \text{ MeV}.$$

A uranium bomb is about as large as a grapefruit and contains about one kilogram of uranium. The quantity of energy liberated when uranium undergoes fission is about 2×10^{10} calories per gram. A rough estimate of the specific heat of the material formed immediately after the fission process must take into account not only the energy needed to provide kinetic energy to all the particles, but also the energy needed to vaporize everything and to ionize everything, in some cases until the atoms are stripped down to their bare nuclei. Instead of the usual value of specific heat of about 1 calorie per gram-degree, we get about 400 calories per gram-degree. The temperature rise ΔT to be expected immediately after explosion (in a time interval from 1 to 5×10^{-8} sec) is therefore

$$\Delta T = \frac{\text{energy liberated per gram}}{\text{specific heat}}$$

$$= \frac{2 \times 10^{10} \text{ cal/gm}}{400 \text{ cal/gm deg}}$$

$$= 5 \times 10^7 \text{ }^\circ\text{K}.$$

At first, the intense radiation (mostly X rays) emitted by this small volume of tremendously hot gas escapes and is absorbed in the surrounding air, thereby raising *it* to incandescence. A spherical mass of hot gas at approximately a uniform temperature throughout is formed in about 1 microsecond and is called the *ball of fire*. The pressure of the gases in the ball of fire starts at about a million atmospheres. The ball of fire starts to expand, compressing the air near it, so that there is a large rise of temperature due to the compression alone. The sudden compression of surrounding layers of air constitutes a shock wave whose front, after about 1 millisecond, coincides with that of the ball of fire. At later times, the shock front travels faster than the ball of fire at a speed of about Mach 14. After about 1 second, the radius of the shock front is about 600 feet whereas that of the ball of fire is only 450 feet.

The ball of fire cools as it expands and as radiation escapes from it. There are so many complicated processes taking place, however, that the temperature change is not monotonic. Fig. 4-17 shows that the temperature decreases to 2000°K in the first hundredth of a second and then increases to 7500°K in the next tenth of a second.

§4-7 **Fusion reactions.** There are two types of nuclear reactions in which large amounts of energy may be liberated. In both, the rest mass of the products is less than the original rest mass. The fission of uranium, already described, is an example of one type. The other involves the combination of two light nuclei to form a nucleus which is more complex, but whose rest mass is less than the sum of the rest masses of the original nuclei. Examples of such energy-liberating reactions are as follows:

$$_1\text{H}^1 + {}_1\text{H}^1 \rightarrow {}_1\text{H}^2 + {}_1\text{e}^0,$$
$$_1\text{H}^2 + {}_1\text{H}^1 \rightarrow {}_2\text{He}^3 + \gamma\text{-radiation},$$
$$_2\text{He}^3 + {}_2\text{He}^3 \rightarrow {}_2\text{He}^4 + {}_1\text{H}^1 + {}_1\text{H}^1.$$

In the first, two protons combine to form a deuteron and a positron. In the second, a proton and a deuteron unite to form the light isotope of helium. For the third reaction to occur, the

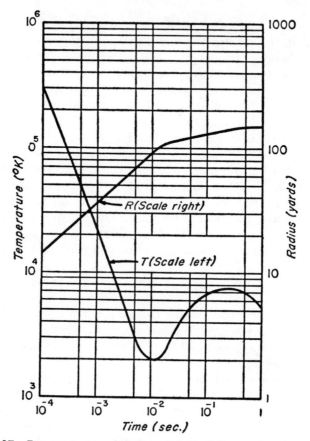

FIG. 4-17   **Temperature and radius of ball of fire as function of time after explosion.**

first two reactions must occur twice, in which case two nuclei of light helium unite to form ordinary helium. These reactions, known as the *proton-proton chain,* are believed to take place in the interior of the sun and also in many stars which are known to be composed mainly of hydrogen. The positrons produced during the first step of the proton-proton chain collide with electrons; annihilation takes place, and their energy is converted into γ-radiation. The net effect of the chain, therefore, is the combination of four hydrogen nuclei into a helium nucleus and γ-radiation. The net amount of energy released may be calculated by subtracting the rest mass of 1 helium atom from the rest mass of 4 hydrogen atoms. The result is 26.7 MeV. In the case of the sun, a gram of its mass contains about 2×10^{23} protons. Hence, if all of these protons were consumed, the energy released would be about 5×10^{10} calories. If the sun were to continue to radiate at its present rate, it would take about 30 billion years to exhaust its supply of protons.

Temperatures of millions of degrees are necessary to initiate the proton-proton chain. A star may achieve such a high temperature by contracting and consequently liberating a large amount of gravitational potential energy. When the temperature gets high enough, the reactions occur, more energy is liberated, and the pressure of the resulting radiation prevents further contraction. Only after most of the hydrogen has been converted into helium will further contraction and an accompanying increase of temperature result. Conditions are then suitable for the formation of heavier elements.

Temperatures and pressures similar to those in the interior of stars may be achieved on earth at the moment of explosion of a uranium fission bomb. If the fission bomb is surrounded by proper proportions of the hydrogen isotopes, these may be caused to combine into helium and liberate still more energy. This combination of uranium and hydrogen is called a "hydrogen bomb."

Attempts are being made at this time all over the world to control the fusion of hydrogen isotopes so as to utilize the resulting energy for peaceful purposes. Two of the reactions being studied are

$$_1H^2 + {}_1H^2 \rightarrow {}_1H^3 + {}_1H^1 + 4 \text{ MeV}, \tag{1}$$
$$_1H^2 + {}_1H^3 \rightarrow {}_2He^4 + {}_0n^1 + 17.6 \text{ MeV}. \tag{2}$$

In the first (known as a D-D reaction), two deuterons combine to form tritium and a proton. In the second (known as a D-T reaction), the tritium nucleus combines with another deuteron to form helium and a neutron. The result of both of these reactions taking place in succession is the liberation of 21.6 MeV of energy.

For two deuterons to come close enough to fuse they must approach each other with high velocity. Such a velocity in a star is the average velocity of a Maxwellian distribution corresponding

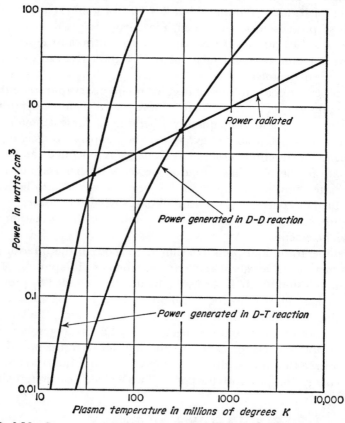

FIG. 4-18 **Power generated versus plasma temperature. Enclrcled points show "ignition temperatures" above which the fusion reaction is self-sustaining. (From Project Sherwood, A. S. Bishop, Addison-Wesley Publishing Co., 1958.)**

to a high temperature. In other words, high velocities in a star are *random, thermal* velocities. There is no reason that the velocity of approach of two deuterons need be thermal. If the velocity is such that the kinetic energy is 1 eV, the temperature of a gas whose molecules *would* have this average kinetic energy is called its *kinetic temperature,* and is given by

$$kT = 1 \text{ eV} = 1.6 \times 10^{-12} \text{ erg,}$$

or

$$T = 11{,}600°\text{K.}$$

If the kinetic energy is 4000 eV, the velocity would equal the average velocity of a Maxwellian distribution associated with a temperature of $4000 \times 11{,}600 = 46{,}400{,}000°$K. At such a kinetic temperature, hydrogen fusion reactions may occur.

The conditions necessary for a hydrogen fusion reaction are approximately achieved in a plasma through which an enormous momentary current is established by discharging huge banks of capacitors. The pinch effect and other external magnetic fields confine the region of high atomic velocities, and a small amount of fusion has been achieved. The energy released, however, thus far has been small compared with the energy required to operate the equipment.

For practical utilization, a fusion reaction must be self-sustaining. The power generated must be greater than the power lost. Since power is lost mainly through radiation, the temperature at which a fusion reaction could become self-sustaining can be obtained by plotting two curves, one of power generated versus temperature and the other of power radiated versus temperature. The point of intersection gives the temperature at which the reaction first becomes self-sustaining. These curves are shown in Fig. 4-18 for both the D-D and the D-T reactions.

5. Beyond Infinity to Negative Temperatures

§5-1 **Subsystems.** At the end of Chap. 2 we learned that the Boltzmann equation enables us to use the concept of temperature in situations where previously we would not have dared. Consider, for example, the mixture of atoms, ions, and electrons inside a tube containing a mixture of gases through which an electrical discharge is maintained, such as a fluorescent lamp. When we touch the outer wall of the tube, it feels slightly warmer than the surrounding air, and we realize that the temperature *of the glass* is about 300°K. But what about the electrons inside? Experiment shows that the distribution of electrons among the energy levels (in simple language, the distribution of velocities) is very similar to an equilibrium distribution given by the Boltzmann equation. It is a tremendous temptation, therefore, to ask, "What numerical value of the temperature must be substituted for T in the Boltzmann equation in order to provide the same distribution of electron velocities that is actually found in the discharge tube?" The answer is found experimentally to be in the neighborhood of 10,000°K! This is called the *electron temperature,* and it refers to a *part* of a composite system, a part that has a sort of equilibrium of its own, and to which a statistical temperature may be ascribed. Such a part is called a *subsystem.*

When can one speak of the temperature of a subsystem? Must one have 100 percent equilibrium? Perfect equilibrium is a sterile delusion. If it existed, nothing would ever happen, and the physicist would be left with a beautiful value of T with which he could do nothing. On the other hand, if a gas is on one side of a partition with a vacuum on the other side and the partition is broken, the gas rushes into the evacuated space with many sorts of phenomena present which are incapable of nice, neat macroscopic description. The states that the gas traverses are hopeless, non-equilibrium states. A liquid pressure gauge would jiggle up and down so violently that a pressure reading would be impossible. A thermometer would register nothing sensible. The very concept of temperature could not be used. This is the other extreme.

The conditions that must exist in order that the concept of temperature be appropriate for a subsystem are simply these:

(1) *The particles of the subsystem must come to equilibrium among themselves quickly;* and

(2) *Equilibrium between the subsystem and the rest of the composite system must take place slowly enough to enable satisfactory measurements to be made.*

The main point is that, in the time interval at the disposal of the experimenter, the manometer liquid must not move about so violently that a reading cannot be taken, and other macroscopic quantities must not fluctuate to a degree that precludes measurement.

If some hot coffee is put into a *dewar flask* (known to picnickers as a thermos bottle), the coffee and the interior walls quickly come to thermal equilibrium at a high temperature, whereas the outside air and the outside walls of the dewar flask come to thermal equilibrium at a much lower temperature. The coffee and the inner part of the flask constitute a subsystem provided that the time for the hot subsystem to come to equilibrium with the cold remainder of the system is so great that one can make measurements on the coffee. Since equilibrium of the coffee alone takes only a few seconds, whereas equilibrium between the coffee and the outside air takes many hours, the concept of temperature is applicable.

The subsystem that is most intriguing in present-day physics consists of the tiny magnets associated with the *nuclear spins* of the ions in a crystal lattice. This is the subsystem whose orientation, when controlled by an external or an internal magnetic field, was used to obtain extremely low temperatures.

The rest of the composite system is the *vibrating crystal lattice.* Here is a peculiar situation: this subsystem is not *spatially* distinct from the rest of the system as are the electrons in a gas discharge or the hot coffee particles in a dewar flask. The nuclear magnetic spins are associated with some of the particles of the crystal lattice itself. This does not prevent us from considering the nuclear spins as a subsystem in its own right, for the nuclear magnets come to equilibrium among themselves in a small fraction of a second due to their interpenetrating magnetic fields, whereas the nuclear magnets may take from 5 to 30 minutes to come to equilibrium with the rest of the crystal lattice.

§5-2   **Negative values of the Kelvin temperature.** Let us recall the original definition of the Kelvin scale of temperature: Two Kelvin temperatures are to each other as the heats transferred during isothermal processes at these temperatures, provided these isothermal processes terminate on the *same* adiabatic curves. If Q and Q_s are the absolute values of the heats transferred at temperatures T and T_s, respectively, the original Kelvin definition provides the relation

$$T = T_s \frac{Q}{Q_s}.$$

If T_s refers to an arbitrary standard, the choice of a number for T_s is also arbitrary. If it is chosen to be negative, then all temperatures would be expressed by negative numbers. Whether T_s is chosen positive or negative, as Q is made smaller and smaller in any unordered way, the limiting value of Q is zero (i.e., the least amount of heat that can be transferred is no heat at all), and therefore *the lowest value of T is zero.* In other words, the lowest temperature is absolute zero, and if negative temperatures have any meaning at all, *they cannot mean temperatures colder than absolute zero!* But what is meant when the Kelvin scale is defined in the usual way with positive numbers?

A clue as to the meaning of negative Kelvin temperatures is provided by the expression for temperature used in statistical thermodynamics,

$$T = \frac{dU}{dS} \text{ (no work).}$$

The most familiar thermodynamic systems, such as a mole of ideal gas or a mole of crystal, have an infinite number of energy levels. As the temperature is raised, more and more atoms are raised to higher energy levels. This requires more and more energy, and results in greater and greater disorder as the atoms are distributed over more and more states. As the energy goes up (positive dU), the entropy also goes up (positive dS), and hence the ratio dU/dS is positive. For T to be negative, *an increase of energy would have to be accompanied by a decrease of entropy!* This obviously cannot take place when a system has an infinite number of energy levels.

Another way of looking at the matter is with the aid of the Boltzmann equation,

$$\frac{n_i}{n_0} = e^{-\frac{u_i - u_0}{kT}}.$$

If the system has an infinite number of energy levels, an increase of temperature produces increased populations of higher and higher energy levels, but no energy level ever gets populated more than the one below it, so that the ratio n_i/n_0 is always less than one, and T is positive. At $T = \infty$, n_i would be equal to n_0, but this would require an infinite amount of energy because of the infinite number of energy levels! Evidently, for T to be negative, n_i would have to be *larger* than n_0, that is, the upper energy levels would have to be populated *more* than the lower ones. This would require even more than infinite energy, which is even more than nonsensical. We conclude, therefore, that in the case of an ordinary system which has an infinite number of energy levels, negative temperatures are an absurdity.

But what about a system which has only a finite number of energy levels? Suppose for the sake of argument that a system were capable of existing in only two energy levels. Let the system consist of N particles and the levels have energies 0 and ϵ. The curve showing the relation between entropy S and energy U is shown in Fig. 5-1. At zero energy, all N atoms are in the lower energy level, which is a state of minimum disorder, or zero entropy. When the two energy levels are equally populated, the energy of the system is $N\epsilon/2$ and there is maximum disorder and hence maximum entropy. If and when all N atoms are in the upper energy level, $U = N\epsilon$, and again we have minimum disorder, or zero entropy. The left half of the curve has a positive slope, and therefore $T \ (= dU/dS)$ is positive. *The right half, with negative slope, is the region of negative temperatures.*

As we start at the origin in Fig. 5-1 and go to the right, we proceed in the direction of increasing energy, increasing hotness,

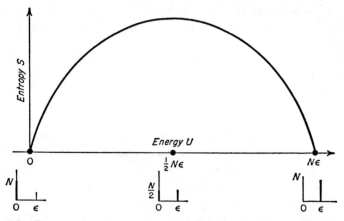

FIG. 5-1   Entropy versus energy curve for a system of *N* particles which can exist in only two energy levels.

and therefore increasing temperature. At the position of maximum entropy where both energy levels are equally populated, the temperature is infinite, for only in this way can

$$\frac{n_1}{n_0} = e^{-\epsilon/kT} = 1.$$

Beyond the maximum, the temperature must be hotter than infinity. Hence, *negative temperatures are hotter than infinity!* If we object to this conclusion, we are really objecting to the original definition of the Kelvin scale. Everything would be much neater if we defined a new quantity, the "negciptemp," $\mathfrak{N}$, equal to the negative reciprocal of the Kelvin temperature; thus

$$\mathfrak{N} = -\frac{1}{T}.$$

When S is plotted against the negciptemp, the resulting curve is shown in Fig. 5-2. Here absolute zero is at an infinite distance to the left, the origin is at $T = \infty$, and negative temperatures are to the right of the origin, with "minus zero" and infinite distance to the right. The fact that the negciptemp of absolute zero is $-\infty$

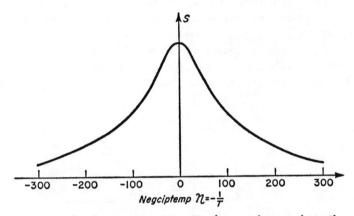

FIG. 5-2 Graph of entropy versus $\mathfrak{N}$, the negciptemp (negative reciprocal of the Kelvin temperature).

is particularly appropriate in view of the third law of thermodynamics. It somehow seems more natural that a temperature of minus infinity should be unattainable.

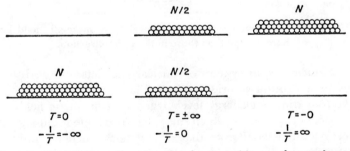

FIG. 5-3 Populations of energy levels at positive and negative temperatures and negcitemps.

The distribution of atoms between only two energy levels is shown in Fig. 5-3 along with the appropriate temperatures and negciptemps.

§5-3   **Achievement of negative temperatures.** To produce negative temperatures we must find a subsystem with the following properties:

(1) The particles must have a *finite* number of energy levels;

(2) The particles must come to equilibrium with one another very rapidly; and

(3) The particles must come to equilibrium with their surroundings slowly enough to enable an experiment to be done.

The first subsystem that was found to satisfy these conditions

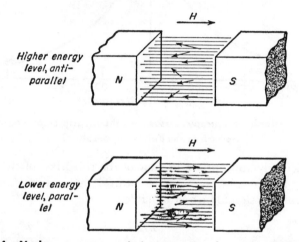

FIG. 5-4   **Nuclear magnets pointing opposite the magnetic field have more energy than those in the same direction.**

is the nuclear spin system, or nuclear magnets, of the lithium ions in a lithium fluoride crystal. These nuclear magnets can have four different energy levels when in a magnetic field. The behavior of particles which may be distributed among four energy levels is similar to that of a system which has only two energy levels. To save words, therefore, let us continue to talk about only two energy levels. In a magnetic field, nuclear magnets may align themselves either in the same direction as the field, as shown in the lower diagram of Fig. 5-4, or opposite the direction of the field, a state of higher energy shown in the upper diagram of Fig. 5-4.

Under ordinary circumstances, there are many fewer nuclear magnets in the upper energy level than in the lower. The subsystem is in equilibrium with itself and also in equilibrium with the rest of the lattice, so that both the subsystem and its surroundings have the same positive temperature. Now suppose we *quickly reverse the direction of the external field*—so quickly, indeed, that the nuclear magnets are unable to follow the change of direction of the field. The large number of nuclear magnets which were formerly in the direction of the field (and in the lower energy level) now find themselves oriented opposite the field, and therefore in the upper energy level! The few nuclear magnets formerly in the upper state are now in the lower one. *There has been a population inversion.* After a small and rapid reshuffling, the nuclear magnets come to equilibrium with one another, and the temperature of the subsystem is negative. After a while (from 5 to 30 minutes), the subsystem cools off and comes to equilibrium with the rest of the lattice and regains its former positive temperature. This beautiful and important experiment was first performed by Purcell and Pound at Harvard University in 1951.

But, how can one be sure that a population inversion has taken place, that is, that there are more particles in the upper energy level than in the lower? The answer is, by measuring the absorption and re-emission of electromagnetic waves whose wavelength λ is connected with the energy difference of the two levels ϵ, by the Planck relation, $\epsilon = hc/\lambda$. When ϵ is the small value associated with Zeeman levels, λ is large compared with visible light, and is in the region of microwaves. When a beam of microwaves with this particular λ is sent through the lithium nuclei, two processes affecting the beam take place:

(1) The beam is *reduced* in intensity by raising nuclei in the lower state to the upper state, the amount of reduction depending on the number of nuclei in the *lower* state. This is ordinary *absorption*.

(2) The beam is *increased* in intensity by causing transitions of nuclei from the upper state to the lower state. A nucleus so lowered emits a quantum of radiation *in phase* with the radiation that forced it down, and the consequent increase in intensity

of radiation is proportional to the number of nuclei in the *upper* state. This is called *stimulated emission*.

When absorption outweighs stimulated emission, the beam is reduced in intensity, and we conclude that there are more particles in the lower state than in the upper. When stimulated emission outweighs absorption, the beam is stronger, and we conclude that there are more particles in the upper state than in the lower. A population inversation therefore enables one to amplify radiation. Subsystems at negative temperatures therefore can produce *Microwave Amplification through Stimulated Emission of Radiation*. The name *maser* comes from the initial letters of the words in this phrase.

§5-4 Thermodynamics at negative temperatures. Classical thermodynamics takes some peculiar twists at negative temperatures, but much remains the same as at positive temperatures. Take, for example, the entropy principle, which states that the sum of all the entropy changes accompanying a natural irreversible process is positive. Suppose that Q units of heat leave a hot reservoir at a temperature, say, of $-40°K$ and enter a colder reservoir at, say, $-100°K$, as shown in Fig. 5-5(a). (Recall that the hottest negative temperature is -0 and the coldest is $-\infty$.) Since heat leaves the hotter reservoir, the entropy change is $-Q/-50$, whereas the entropy change of the colder reservoir is $+Q/-100$. The total entropy change is

$$\frac{-Q}{-50} + \frac{Q}{-100} = \frac{Q}{100},$$

which is positive, just as it is with positive temperatures.

Everyone knows that the outside air, the oceans, the rivers, etc., possess a tremendous store of internal energy which is there for the asking. You don't have to buy fuel and an expensive furnace for burning the fuel. All you have to do is to extract some of the internal energy from the air or from the ocean, put it into a heat engine, and convert it into work. The work may be utilized in generating electric power, or in driving a ship across the ocean, or in propelling a locomotive across the land. It was recognized by the great founders of thermodynamics, Kelvin, Clausius, and Planck, that, however great a boon to mankind such an energy conversion would be, it is quite contrary to experience, so that it is an accepted statement of the second law of thermodynamics (the *Kelvin-Planck* statement) that it is impossible to convert heat completely into work without producing changes somewhere else.

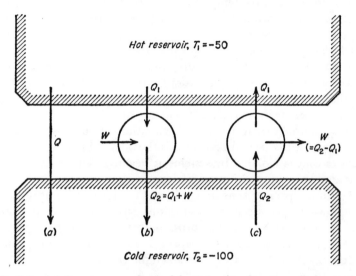

FIG. 5-5 **(a) Spontaneous flow of heat in the direction of decreasing temperature. (b) a costly device for doing something that requires no device. (c) A heat engine that could be used to convert heat Q₂ − Q₁ from the cold reservoir completely into work.**

In Fig. 5-5(b), an attempt is shown to imitate the way a heat engine behaves, but at negative temperatures. Since, by definition of the Kelvin scale,

$$\frac{Q_1}{Q_2} = \frac{T_1}{T_2} = \frac{-50}{-100} = \frac{1}{2},$$

when Q_1 units of heat leave the hotter reservoir, *twice as much* heat must enter the colder one. Therefore, instead of work W being done *by* the engine, work would have to be done *on* the engine in order not to violate the principle of the conservation of energy. But the device depicted in Fig. 5-5(b) is awfully silly because it is an expensive gadget for doing a job that requires no device at all. If all you want is to push heat into a cold reservoir, it is sufficient merely to allow Q_1 to flow *naturally* from the hot to the cold reservoir.

To get W units of work *out* of a heat engine operating between reservoirs at negative temperatures, you would have to make use of the device shown in Fig. 5-5(c), where Q_2 units of heat are taken *from* the cold reservoir (as though it were a refrigerator). Then a *smaller* quantity Q_1 would go into the hotter reservoir and the rest would be available for work. But the hot reservoir could be dispensed with, for the Q_1 units of heat would naturally flow back to the colder reservoir. The net result would be that $Q_2 - Q_1$ units of heat were extracted from the *colder* reservoir and converted completely into work, in violation of the Kelvin-Planck statement of the second law. This is the only principle of classical physics that is violated by systems at negative temperatures—but it is an important and interesting one.

Up to the present time, the only real use for systems at negative temperatures has been in the rapidly expanding field of masers and lasers. Perhaps, in the future, experiments on heat engines and refrigerators will be performed at negative temperatures. Then, it will really be fun to be an engineer.

Bibliography

GENERAL

M. W. Zemansky, *Heat and Thermodynamics* (McGraw-Hill Book Co., 1957).

D. K. C. MacDonald, *Introductory Statistical Mechanics for Physicists* (John Wiley & Sons, 1963).

LOW TEMPERATURES

K. Mendelssohn, *Cryophysics* (Interscience Publishers, 1960).

F. E. Hoare, L. C. Jackson, and N. Kurti, *Experimental Cryophysics* (Butterworth & Co., 1961).

J. Wilks, *The Third Law of Thermodynamics* (Oxford University Press, 1961).

E. M. Lifshitz, "Superfluidity," Sci. Amer., June 1958.

K. Mendelssohn, "Superfluids," Science **127**, 215 (1958).

N. Kurti, "Nuclear Orientation and Cooling," Physics Today, March 1958.

H. A. Boorse, "Some Experimental Aspects of Superconductivity," Am. J. Phys. **27**, 47 (1959).

"Superconductivity Maintained under High Magnetic Fields," N. B. S. Tech. News Bull., Jan. 1962.

HIGH TEMPERATURES

F. G. Brickwedde, editor, *Temperature,* Vol. III, Part I, with C. M. Herzfeld, editor-in-chief (Reinhold Pub. Corp., 1962).

H. Fischer and L. C. Mansur, editors, *Conference on Extremely High Temperature* (John Wiley & Sons, 1958).

M. McChesney, "Shock Waves and High Temperatures," Sci. Amer., Feb. 1963.

NEGATIVE TEMPERATURES

N. F. Ramsey, "Thermodynamics and Statistical Mechanics at Negative Absolute Temperatures," Phys. Rev. **103**, 20 (1956).

E. M. Purcell and R. V. Pound, "A Nuclear Spin System at Negative Temperature," Phys. Rev. **81**, 279 (1951).

Index

Absolute zero, 23, 114
Absorption, 119
Absorptivity, 80
Adiabatic demagnetization, 50
Adiabatic process, 18
Adiabatic wall, 2
Allen-Bradley radio resistors, 43
Ambler, 67
Arc, 86
Augmentation, electrical, 79

Bardeen, 58
Bishop, 110
Blackbody, 81
Blaisse, 73
Bleakney, 104
Bleaney, 67
Boiling point, 9
Boltzmann, 31
Boltzmann's constant, 31
Boltzmann's equation, 35
Boorse, 123
Brickwedde, 123
Brillouin, 48
Brute force method, 66

Cailletet, 75
Celsius temperature scale, 17
Cerium magnesium nitrate, 58, 73
Chromium potassium alum, 54, 57
Clausius, 120
Clement, 43
Collins, 40, 75
Collins helium liquefier, 40, 41
Conservation of parity, 69
Cooper, 58
Coordinate, state, 1

Curie's constant, 48
Curie's law, 48

Daunt, 64
Debye, 50
De Klerk, 54, 55, 56
Dewar, 75
Dewar flask, 113
Diamagnetic, 59
Diathermic wall, 3
Diffraction, 92
Doppler breadth, 95
Doppler effect, 93
Doppler temperature, 99

Einstein A coefficient, 98
Electrical augmentation, 79
Electron temperature, 112
Emission, stimulated, 120
Emittance, radiant, 80
Energy, 25
Entropy, 26, 31
Excitation temperature, 99

Fahrenheit temperature scale, 17
Film creep, 44
First law of thermodynamics, 26
Fischer, 123
Fission of uranium, 106
Fixed points, 14, 42
Flames, 78
Fountain effect, 44
Fowler, 74
Fusion, 107

Gas thermometer, 8, 13, 23, 41
Germanium thermometer, 43

Giauque, 50, 75
Gorter, 66
Griffin, 95
Guggenheim, 74

Heat, 16
Heat exchanger, 39
Heer, 64
Henry, 48
Hill, 98
Hoare, 123
Hudson, 67

Interference fringes, 101
Interferometer, 102
International temperature scale, 15
Inversion, 38
Ionization, 89
Ionization temperature, 99
Irreversible process, 28
Isentropic, 27
Isothermal process, 18

Jackson, 123
Joule-Kelvin effect, 37
Joule-Thomson effect, 37

Kamerlingh-Onnes, 58, 75
Kapitza, 75
Kelvin, 120
Kelvin temperature scale, 17, 21, 114
Kinetic temperature, 111
Kirk, 75
Kurti, 56, 70, 71, 75, 123

Lambda point, 44
Law of mass action, 88
Lee, 67
Lifshitz, 123
Liquefying a gas, 39

MacDonald, 123

MacDougall, 50, 75
Mach number, 101
Mach-Zehnder interferometer, 102
Magnetic induction, 59
Magnetic moment, 48
Magnetic refrigerator, 64
Magnetic temperature, 54
Magnetization, isothermal, 19, 52
    adiabatic, 20, 52
Mansur, 123
Maser, 120
Mass action, law of, 88
Matthias, 59
McChesney, 105, 123
McNally, 95
Meissner, 60
Melting point, 9
Mendelssohn, 123

Negative Kelvin temperature, 114
Negciptemp, 116
Nernst, 76
Niobium-tin compound, 59, 61
Nuclear fission, 106
Nuclear spins, 113

Olzewski, 75
Optical interferometer, 102
Optical pyrometer, 7, 85

Paramagnetic salt, 18, 46
Parity, conservation of, 69
Permeability, 59
Phase equilibrium, 9
Pinch effect, 87
Plasma, 86
Planck, 82, 120
Planck's constant, 84
Planck's radiation law, 79, 85
Plasma jet, 87
Plasma torch, 87
Plumb, 43
Polarization, 64
Populations of states, 30
Porous plug experiment, 37